ACKNOWLED(

For their much valued support, advice and ~~~~~~~~, and for reviewing the basic manuscript of this volume, I give my utmost, sincere thanks to:

Diane Boyd, Lu Carbyn, Steven H. Fritts, Jim Hammill, Rick McIntyre, L. David Mech, Mary Ortiz, Dennis Senger, Will Waddell and Adrian Wydeven.

For support in the past, I thank:

Tracy Brooks, Wendy Brown, Colleen Clark, Jennifer Gilbreath, Maureen Greeley, Dan Groebner, Jeff Haas, Bobbie Holaday, Jack Laufer, David Parsons, Bill Paul, Rolf O. Peterson, Sharon Rose, Lori Schmidt, Monty Sloan and Janice Templeton.

For reviewing the basic manuscript in the U.K., my thanks to:

Leonard Pearce, Victor Watkins and Trevor Wheeler.

Special thanks to Leonard Pearce and to Richard Sheen for their offer of support in the world of publishing.

Finally, a very big thank you to Carole Stevenson, for helping to make a dream become a reality.

© November 1998
Redman. I. Savage Freedom.........The World Of The Wolf

Front Cover ~ Photograph by L. David Mech

Photographic Credits

page 5	L David Mech
page 7	Diane Boyd
page 10	Don Adams
page 12 (top)	Tracy Ane
page 12 (bottom)	Tracy Ane
page 23	Ian Redman
page 26 (top)	Ian Redman
page 26 (bottom)	Ian Redman
page 31	Ian Redman
page 32	Ian Redman
page 33	Tracy Ane
page 39	L David Mech
page 40 (top)	L N Carbyn
page 40 (bottom)	Diane Boyd
page 42	L N Carbyn
page 45	Tracy Ane
page 55 (top)	Ian Redman
page 55 (bottom)	Ian Redman
page 58	L David Mech
page 67 (top)	Point Defiance Zoo and Aquarium
page 67 (bottom)	Charles Bergman
page 70	Point Defiance Zoo and Aquarium
page 72	L David Mech
page 73	L. David Mech
page 75	Carole Stevenson
page 78	L N Carbyn
page 79	Ian Redman

DEDICATION

For my mother and father, Dorothy and Ernest Redman
"thank you both for everything".

And

Carole and Niel Stevenson and Geoff Davies.

"The very best of friends can only be counted on one hand."

SAVAGE FREEDOM ~ THE WORLD OF THE WOLF

Contents

Foreword by Dr. Diane Boyd	page 6
Introduction by Ian Redman	page 8

Chapter One
The World of the Wolf .... page 13

Chapter Two
Minnesota Magic.....*Radio tracking grey wolves in Northern Minnesota* .... page 22

Chapter Three
Savage Freedom......*Wolves on the hunt* .... page 34

Chapter Four
A Howling Over Russia....*Tracking wolves the hard way* .... page 51

Chapter Five
Wolves and Ravens.....*Spirits of the forest and black thunderbolts* .... page 59

Chapter Six
The Red Wolf.....*Living on the edge* .... page 64

Chapter Seven
Man's Best Friend.......*The wolf* .... page 71

Conclusion — *Wolves, Humans............and the future* .... page 82

Bibliography .... page 84

Appendix i: About WOLF HELP .... page 88

Appendix ii: International wolf conservation .... page 91

About the Author .... page 92

FOREWORD

By Dr. Diane Boyd
Wildlife Biologist ~ U. S. Fish & Wildlife Service

Dozens of books have been written about wolves in the past fifty years. The recent exponential growth in "wolf books" has corresponded with the increasing popularity of wolves. Most of these books were written as documentaries, scientific treatments, historic recollections or coffee table books with exquisite photographs and little text. What has been lacking in these efforts is the story of how a non-scientist turns his impassioned dream of studying wolves into a fulfilling reality, on his own terms. Herein, Ian Redman describes that accomplishment.

When Ian calls me, the conversation always begins with a hearty, bubbling "Hello Diane, it's Ian here! Just thought I'd ring you up to tell you about......" From there our conversation jogs through avenues of wolf conservation, biology, research and public education and concludes with a thorough travelogue of places and audiences to which Ian has recently delivered his impassioned Wolf Help lectures. His enthusiasm prevails throughout this book, Savage Freedom : The World of the Wolf. Ian provides a British perspective on wolf biology and his path of personal involvement, through the eyes of an ardent wolf educator and an adventurer.

This easy to read volume is the story of a man who has spent over a decade working tirelessly for public education about wolves. Ian is the first to admit that he is not a scientifically trained researcher, but rather has come into the wolf world through his own sincere passion for the wolf and its conservation. He is compelled to spread his good words about wolves to a wide variety of audiences. Ian is a living example of how satisfaction comes to those who pursue their dreams.

Ian first became interested in wolves whilst watching a television documentary on arctic wolves. Since that time, he has contacted numerous wolf biologists with his questions, trying to obtain accurate information for his popular wolf education programmes. In both his presentations and this book he discusses basic wolf biology in layman's terms. Following this basic introduction he takes us on an imaginative journey to rendezvous with wolf biologists across North America and Russia. He explores the connections between wolves and their prey, wolves and ravens, wolves and dogs and the often uneasy relationship of wolves and humans. The latter topic is the heart of Ian's outreach programmes, because change comes about through education.

The challenges involving wolf and human coexistence are considerable. Paul Errington summed it up in his book 'Of Predation and Life' when he stated "Of all the native biological constituents of a northern wilderness scene, I should say that the wolves present the greatest test of human wisdom and good intentions". Ian, with his enthusiastic embrace of wolf charisma, is attempting to create tolerance of wolves that will enhance survival of this superb predator.

INTRODUCTION

The three Mexican wolves looked me straight in the eye. What would be their next move? Which way would they run? Left or right? I took a firm hold of my large capture net as I decided to intercept the yearling wolf standing between the other two. The next few minutes became chaotic as my good friend, and Wolf Help colleague Niel Stevenson, and I tried our best to capture these beautiful wild dogs in our nets.

At that time, in October 1997, the two of us were working with staff members from Wolf Haven International and two Red wolf biologists. Our task was to capture a total of eleven captive-bred Mexican wolves and to assist with their important veterinary care, a step that would eventually lead to the release of several of them into the wilds of Arizona just a few months later.

For both Niel and myself, this experience was a dream come true and it was certainly a highlight of my years of wolf research and conservation.

My interest in wolves started quite by accident back in 1989. I remember watching a wildlife documentary featuring a biologist by the name of L. David Mech and a cameraman named Jim Brandenburg. Both men took their audience, including myself, deep into the world of the Arctic wolf on Ellesmere Island, approximately 500 miles from the North Pole. As I continued to watch this fascinating programme, I began to realise just how incredible wolves really are. These beautiful wild dogs which inhabit the Northern Hemisphere are so intelligent, social and, in many ways, courageous. Theirs is a complex society requiring a vast amount of study. As the credits were shown at the end of this engrossing documentary, I knew then that I wanted to immerse myself completely into the world of the wolf.

I have never looked back since......

All of my wolf studies have been, and will continue to be, in my free time. This is because I have a career in the food industry in the British Isles. To say, however, that I study wolves as a 'hobby' is highly inappropriate. I absolutely 'live' for these animals.

In 1992 I travelled to Minnesota, and to Russia in 1993, for wolf research, returning to Minnesota in 1995. Over the years, I have developed special friendships with many of North America's finest wolf research biologists, all of whom support my quest for information about wolves.

In 1993 I met Carole and Niel Stevenson. The three of us immediately became good friends and, within a few months, we began to travel to schools, colleges and other educational establishments, giving our very first photographic slide presentations on both grey and red wolves.

In Carole and Niel I found two special people who both shared my deep respect and passion for wolves.

As time moved on, our friendship developed even further. Now, the three of us were enjoying a continuation of great support, which I had already frequently received, from wolf biologists in both America and Canada.
In 1995 Carole, Niel and I decided to form 'Wolf Help' (Helping Education for Lobo Preservation) and to take our enthusiasm and knowledge of wolves even further as we travelled the length and breadth of the British Isles, speaking to both young and old people alike. Again, this was all in our free time.

Carole is responsible for the creation of the Wolf Help display boards and works tirelessly on her computer to produce beautiful and factual eye-catching display material.

Niel is perfect as the Wolf Help projectionist, helping to enhance my presentations with his knowledge of slide projectors and lenses.

As the main researcher and speaker for 'Wolf Help', I have now given presentations to literally thousands of people on wolf society, behaviour, biology, taxonomy and wolf research. The three of us have also been guests at Wolf Haven International in the United States where I gave several presentations.

However, I am not a wildlife research biologist. I have no initials after my name and no degrees of any kind. I have taught myself about wolves, and I am very proud of my achievements.

As far as I am concerned, you do not have to be an 'expert' on wolves to make a difference, and to help in wolf conservation issues, but what you do need is dedication, commitment and, most of all, enthusiasm.

Savage Freedom......The World Of The Wolf is the culmination of my years of wolf research and conservation. I have written this book not only to give a detailed view of wolf society, but to give the reader a taste of what it is like to experience the power of the wolf in the wild.

I do hope that you will enjoy reading this book as much as I have enjoyed writing it. I also hope that it will help you to venture even further into wolf conservation and education in any way you can.

If these aims have been achieved, then I will be very happy, and this book will have served its purpose.

Ian Redman

Wolf Researcher/Conservationist and Co-founder (Wolf Help)

Ian Redman right with Wolf Help colleague Niel Stevenson participating in the Mexican Wolf Recovery Programme

"Once this way is accepted by our kind, we should rejoice in the knowledge that there is still, in wilderness, a realm of savage freedom".

Ian Redman
Chapter Three - Wolves on the hunt

CHAPTER 1

The World Of The Wolf

On open tundra, they can hear up to ten miles away…….. In boreal forest, they can hear up to six miles away……..They can smell their prey up to a mile and a half away……..Their society is much akin to human society…….and their basic forms of communication are far more advanced than ours.

They…….are grey wolves.

What is it about their power? What is it about their beauty, their strength and their intelligence that brings the grey wolf to the very forefront of the human imagination? When you 'cry wolf', you actually cry out for humankind, for the wolf has always been the natural scapegoat for mankind's inhumanity to man, and for our inhumanity to the other animal species on this planet. Now, as we live in a world of technological pandemonium the grey wolf can still, in many ways, be found deep inside the human psyche.

So, why do grey wolves fascinate humans? Indeed, there is no other animal on earth that holds such power in the human imagination. From Little Red Riding Hood to the werewolf legends, for hundreds of years in European culture, the wolf has been seen as a malignant, evil animal. In comparison, over in North America, many of the native peoples of this continent respected this wild dog as a spirit guide within the realms of Mother Nature itself, then with the arrival of the European settlers, once again, came hatred and prejudice against the grey wolf.

The real wolf is, in fact, nothing like the blood-curdling beast of waste and desolation that Europeans have always imagined. The real wolf, out in the great wild areas of the Northern Hemisphere is a shy, highly intelligent and social animal which should demand our respect and understanding.

In many ways the wolf is also a very 'human' animal, as several aspects of its society and behaviour convey a deep and significant meaning to the role of nature itself, and to our role on planet Earth.

So, let us start our journey into wolf society; and let us discover the truth about this beautiful wild dog.

Welcome to the world of the grey wolf………..

What is a wolf?

The grey wolf is the largest of the Canidae, this being the scientific collective name for all the worlds' dogs, whether wild or domesticated. The wolf itself once inhabited all the main areas of the Northern Hemisphere including Europe, North America, Japan and Asia. The only exception to these habitat areas were vast deserts and high mountain ranges. Now, of course, the wolf's distribution is much more restricted due to human persecution and habitat loss, as wolves now live in less than one half of their former habitat range.

The grey wolf's main scientific Latin name is Canis lupus, but this genus/species classification encompasses an array of scientifically recognised sub-species of grey wolf, distributed throughout the Northern Hemisphere. Earlier this century the North American grey wolf itself was divided into twenty-four sub-species. However, recent taxonomic investigations have suggested re-classification of these findings and it is now generally agreed by wolf research biologists the world over, that the main sub-species of grey wolf inhabiting the Northern Hemisphere are as follows:

North America

Canis lupus arctos	Arctic Wolf
Canis lupus occidentalis	Alaskan 'Sub' Canadian Wolf
Canis lupus nubilus	Great Plains Wolf
Canis lupus lycaon	Eastern Timber Wolf
Canis lupus baileyi	Mexican Wolf

Europe

Canis lupus lupus	European Wolf
Canis lupus albus	Eurasian Tundra Wolf
Canis lupus signatus	Iberian Wolf

Asia

Canis lupus pallipes	Asian wolf
Canis lupus laniger	Mongolian/Chinese Wolf

Middle East

Canis lupus arabs	Arabian Wolf

Sizes of grey wolves can vary a great deal depending on climate and type of habitat. The average standing height for a male grey wolf can be anything from 26 to 34 inches at the shoulder. Length can also vary from between 5 to 6.5 feet from nose to tail tip. Weights can likewise vary, the average weight of a male grey wolf being anything from 70 to 120 pounds, however, the smallest of the grey wolf sub-species (Canis lupus arabs) can weigh as little as 30 pounds. Moving further North, into Alaska and Canada, wolves tend to be very large, weighing up to 140 pounds or more, then size decreases again as wolves extend into harsh arctic regions.

Life in the pack

Wolves are highly social canids and, in general, their strength for survival in the wild comes from being part of the pack. The grey wolf pack is a family unit consisting of the leaders, older family members, juveniles and pups. The pack itself is a unification of intelligence, tolerance, understanding and co-operative hunting instinct, as every wolf within the pack's social structure will play a vital role, on a day to day basis, for both family matters and the hunt. The actual size of a wolf pack will vary depending on size of territory and prey availability. An average wolf pack size will consist of between four to twelve wolves, but occasionally, packs have been known to consist of up to between 20 to 30 wolves. All wolves within a pack adhere to strict rules regarding hierarchy, and this system is reinforced constantly, thus the pack will have its own leaders. These are called the alpha animals and will consist of a male and a female. In general, the alphas will be the only breeding pair within the pack and they will probably be the strongest of the wolves, and will have pronounced leadership tendencies. All the other wolves within the pack will show great respect to the alphas.

Just below the rank of the alphas will be the beta wolf (or wolves). The beta(s) can be compared to the lieutenants of the wolf world and may be closely related to the alphas, such as brother or sister. The rest of the pack will consist of adults, juveniles, pups and the lowly omega wolf. The omega wolf may be male or female, and is generally the scapegoat of wolf society. This is because the omega, for some reason, does not fit in properly to its own packs' system and will, therefore, be often bullied, chastised and harassed by its own adult pack mates. Life from then on starts to become difficult, and sometimes intolerable for the omega, as it lives on the fringe of pack life, including being the last to feed at the kill. Finally, the omega wolf may become so frustrated with this way of life, that it will leave its own pack permanently. Once the omega decides to leave, it will travel far, carefully avoiding other packs' territories to look for a new mate, thus forming the nucleus for another wolf pack. These animals may also live on the fringe of their former pack's territory existing on leftovers. However, omegas are not necessarily the primary dispersers from the pack, as frustrated betas, yearlings and two-year olds may also disperse at times.

Mating and reproduction

Unlike the domestic dog which comes into estrus twice a year, the grey wolf only comes into estrus once a year, and depending on latitude of habitat, this will vary from between January to early May. In general, it is only the alpha pair that will mate within the pack; the alphas will actively suppress the mating instincts of their sexually mature pack mates at this time by utilising threatening gestures and outward signs of aggression. However, there are occasional exceptions to this, especially in areas of 'high prey density'. This is an area where wolves will have a higher than average ratio of prey species compared to the number of wolves in their pack, and where unclaimed space exists. This will usually ensure a higher success rate on the hunt and will, therefore, bring about a lower mortality rate for the pups in the pack. In this situation, it is possible that the alpha male will mate, or even that the beta male will mate with another female within the pack.

Grey wolves generally reach sexual maturity at 22 months of age, however, there can on occasions be exceptions to this as captive wolves have been known to reach sexual maturity at 10 months of age, and wild wolves at 34 or 46 months. The mating season is also the period of greatest strife within the pack, sometimes causing lower ranking wolves to disperse. During the weeks before mating the alpha animals will become particularly affectionate to each other, then as copulation takes place they will stay locked together by their genitalia for approximately 30 minutes. This is commonly called 'the tie' and will help ensure successful fertilisation of the female.

After a gestation period of approximately 63 days, the alpha female will give birth to a litter of around 4 to 8 pups, in an excavated den site. The den itself will have been dug out by the female, possibly some months before and will, most probably, be in an elevated, well drained position near to water. Most dens are dug into the soil extending back five to fifteen feet, however, hollow logs, old beaver lodges, rock caves or a simple surface bed above ground may be used for a den site. It is also worth noting that, in colder climates, wolves generally prefer a south-facing area for their dens.

When grey wolf pups are new born, they look like any other canid puppies. They will be dark coloured, weigh approximately 1 pound and will be born blind and deaf. At this stage of life they are totally helpless and are fully dependent on their mother. However, it is the first three to four weeks of life that begins the packs' socialisation process for the growing pups. As the alpha female, being the caring, affectionate mother that she is, keeps her offspring close to her, in the warmth of the den, the pups will start to feel the closeness of the pack, and the feeling of family cohesion.

After approximately 4 to 6 weeks, the pups will grow quickly and their sight and hearing will have developed. At this time all wolf puppies eyes are blue, but after a few more weeks of growth, their eyes will develop the more familiar colour that is generally associated with wolves, which is amber.

At around 4 weeks of age, the pups will start to utilise their incredible senses and will begin to show themselves outside the den. Exploration and play are two vital aspects of life for young grey wolf puppies and even at the age of 8 weeks it has been known for wolf pups to venture out well over a mile away from the den, but always under the constant supervision of the adults.

If there is one aspect of wolf society that has greatly impressed wildlife research biologists, it is the care and affection shown by all the members of the pack when taking care of the pups. My years of wolf research have shown me that, in my opinion, it would be difficult to find a more attentive and caring family group, than a wolf pack. The care and education of the young pups is of paramount importance to all the adults, as this secures the future of the pack.

Once the pups have reached the age of 8 to 10 weeks the alpha animals may then decide to leave the natal den and to move the entire pack to what is generally called a 'rendezvous site'. This site may not be anything special as far as the wolves are concerned, as it may be just a meadow or other open area. Here the wolves will feel safe to continue their daily existence and the pups will continue growing, being less bothered by the parasites that, by this time, may have infested the original den. Rendezvous sites are a critical component of wolf habitat, where pups will grow to near adult size, learn to play and hunt together and build the essential family bonds that are the focal point of wolf society. Once late Autumn (Fall) arrives the pups will be large enough to join the adults on regular hunts and permanent rendezvous sites are then no longer occupied as the pack will continue its existence following their prey within their territorial boundary.

After approximately the fifth week of age, the mother wolf will have begun to wean her offspring. The pups will now be fed half-digested meat, which the adults will bring back from the hunt and regurgitate, after specific stimulation from the pups. These stimuli will include licking and nuzzling of the adults' facial and nasal area. However, some adults will regurgitate automatically, knowing instinctively that the pups need feeding. This process helps the pups by familiarising their digestive systems with prey at hand.

By the end of their first year, the pups will have generally reached their full body size, although they will likely gain some extra weight in the second year. However, the most obvious indicator of adult growth will be the size of the pup's leg bones, which will have now grown to maximum length.

At this important time of life, the pups will have now become yearlings and will be ready to join their pack mates in the rigours of life ahead.

Communication.....The language of wolf society

Many people, still retain their first images of wolves as large wild dogs howling at the moon. Of course, the image of a wolf, or of several wolves howling, is a very potent one, but wolves do not howl at the moon. The howling of wolves is just one major part of a vital and important aspect of wolf society and behaviour namely communication.

So how do wolves communicate?

As I previously mentioned, the grey wolf's basic forms of communication are, in many ways, far more advanced than ours. Many aspects of the wolf's senses and body movements are utilised in the role of communication; these include the use of body language, howling and scent marking.

Body language is perhaps the most important aspect of communication in wolf society. All wolves use many parts of their body to convey messages, such as their eyes, ears, teeth, tail and hair. From the highest-ranking alpha wolves, who at times will use very forceful and intimidating body language, to the pups and the omega wolf, both utilising totally submissive body postures. It is also common for alpha wolves to show dominance over the other pack members by snarling, growling, showing their teeth, raising their tails erect and pushing their ears slightly forward. Fear aspects are also communicated by body language, with a tail tucked firmly between the rear legs, ears dropped backwards, and body posture bent low in a 'non-threatening' way. It is a fact that all lower ranking wolves within a pack will show submissive posturing and body language to the alpha animals on many occasions. In general, wolves use strong body language to help avoid actual fighting and ensuing injury. Their dental weaponry is usually reserved for killing prey, not fellow pack members.

Howling is another important aspect in every day communication and is also for conveying various types of messages within the pack, and to other packs. Wolves generally feel an impulse to howl either early morning, late evening or after dark. However, wolves will howl during most times of the day, given certain circumstance and stimuli. Generally, the howling combines a social function with, what I call a defensive 'talk mode'. Most wolves will feel an automatic compulsion to join in, if one wolf starts to howl within the pack. Howling, to a grey wolf is another way of bonding the pack and of feeling a certain kinship among their respective family members, but wolves will also howl for other reasons. Howling defines territory and sends a message to neighbouring wolf packs or to lone wolves to 'keep out'. Wolves will also use howling to stimulate excitement and enthusiasm both before and after the hunt. Lone wolves that have dispersed and left the pack for a while, will seek out their fellow pack mates upon their return by howling, once they have reached their home territory. Dispersing wolves will also howl in the hope of finding and attracting a mate once they have permanently left their natal pack. However, this behaviour will not take place in or around another pack's territory for fear of the wolf drawing attention to its presence. A wolf's howl can carry a great distance, especially at night and wolves also have an uncanny aptitude for reverberating different pitches and tones of their howls so that, from a distance away, a pack of six wolves may sound like a pack of nine or ten. I always remember my first howling session in Minnesota in 1992, when I howled to a pack in the Superior National Forest. At the time of the reply to my human howling, it seemed that there were probably six or seven wolves in the pack, but I later discovered that I had been howling to a small pack of four.

Wolves will also create other sounds for communication. Grunting, whining, squeaking, yelping and barking are all communicative aspects of wolf society. The barking of wolves in particular is used as a major warning signal to other pack members.

Scent marking is also another form of communication in wolf society and is generally used to send messages to other wolves outside a particular territory. When wolves claim territory, they claim a piece of land, which will become their hunting and living area. Sizes of these territories can vary from 20 to 1,000 plus square miles depending on total population size and abundance of prey. A typical wolf pack's territory will measure approximately 100 to 300 square miles. In general, the alpha animals will scent mark all around their given territory in two ways. The most important type of scent marking is called 'raised leg urination', where the alphas will raise one of their rear legs and spray a mixture of urine and scent onto items such as trees or rocks. The scent itself comes from a gland in the anal area. The alphas will also re-affirm their territory and dominance by scraping their paws on the ground. This then deposits a scent from glands in the paws.

Scent marking is designed to send messages not just to other wolves, but to any other animal which may cause a threat to a given territory.

The need for dispersal

Most wolves, at some time in their lives, will feel the need to leave their family and to travel. This is commonly known as dispersal and, in most cases, will be evident in young wolves that have reached sexual maturity. Wolves in larger packs will generally feel a compulsion to disperse during the mating season itself, or just thereafter, when pups have been born and difficulties may arise from having too many mouths to feed in the coming weeks ahead.

Wolves that eventually disperse may travel up to several hundred miles, quite possibly in search of a mate. Sometimes they will be successful, sometimes not. It is certainly not uncommon for dispersed wolves to return to their natal pack, weeks or even months after leaving, only to be welcomed back by their pack mates.

The longest dispersal so far recorded was a radio-collared wolf being studied by Dr. Steven H. Fritts in the Superior National Forest in Northern Minnesota. This wolf was radio-tracked whist on the move between 25 July to 29 December 1980, before it lost its radio collar. The wolf was later located in late 1981 in an area of Saskatchewan in Canada. At this time it was acknowledged that the wolf had travelled 886 kilometres.

Mortality........A question of life and death

In the world of the wolf, life itself can be very arduous indeed. Not only do wild wolves have to deal with the hardships of the hunt in which, on occasions, they may be severely injured or killed, they also have to deal with human attitudes and threats and various types of diseases and maladies.

The average lifespan for wild wolves is, most probably, up to five years of age and various factors will play a major part in deciding a wolf's mortality. In general, the main causes of mortality for wild wolves are death from other wolves due to territorial conflict and the starvation of pups. Wild wolves can also suffer from many types of disease and parasitic infection. Canine parvovirus, distemper, sarcoptic mange, lyme disease and heartworm are all problems for wolves in the wild, so it is important to understand that life-threatening situations for wolves do not just come from the buffeting hoof of a prey animal.

The secrets of pack life

The world of the grey wolf is indeed fascinating and exciting, but it is also a complex world. After many years of wolf research on both captive and wild wolves, the secrets of pack life can now be better understood and appreciated. Only through this appreciation can we, as human beings, offer our understanding and acceptance of this intelligent and highly social wild dog. To have entered the world of the grey wolf is to have entered something very special indeed.

CHAPTER 2

Minnesota Magic
Radio Tracking Grey Wolves in Northern Minnesota

Imagine the scene.............

An eleven-year-old boy looks across the campfire at his father. He smiles warmly, for every weekend was the same, a wonderful adventure away from the bustling metropolis of Minneapolis. The sky is golden-orange as the sun lazes away the last hour of daylight and, in the east, a half-moon crescent bares down on the campers. The complex zigzagging of bats has caught the boy's attention as his father smells the meal cooking in the pot. Slowly, the sun begins to set and they gaze at the lake now lit with a dazzling array of colour. "Nothing else could be better" thinks the boy's father, "absolutely nothing".

Suddenly, the boy turns quickly and yells "look pa.........look!", both sit tight still as six white tailed deer dart from out of the undergrowth. The deer are bucking and jumping up and down by the lakeside, as if in sheer panic. The simmering meal is forgotten and the two spectators stare in amazement as, out of the woods, two silvery-grey shapes appear. Large wild canids, moving slowly, intensely and with great purpose. They make no sound and have their concentration set on the panicking deer in front of them. It is then that the largest animal turns quickly and gazes at the two humans. Both the canids sniff the air, the larger whines, then moves slowly forward towards the campers. Now, just fifteen metres away, two pairs of steel grey eyes make contact with two pairs of human eyes. The boy's father says quietly "at last..........". The canids shrink back, look over at the panicking deer, whine softly to each other, then vanish into the forest. The boy turns to his father in hasty excitement "pa, they were.......". "Yes son" he says, smiling, "they were grey wolves".

* * * * * * * * * * * * * *

After twenty-four hours of continuous travel, utilising four different airports, I was nearing my journey's end, and Jeff, my driver from the International Wolf Center caught my imagination, setting it alight as he drove me northwards, on the last two-hour stretch of my journey into wolf country. It was the sixth of June, 1992, and I was heading right into the heart of Minnesota's wilderness, in the Superior National Forest. Jeff's story of how he and his son had their first encounter with grey wolves, was a wonderful introduction to the land of

ten thousand lakes. A land which is now home to over two thousand, two hundred wolves of the 'Nubilus' sub species of grey wolf, more formerly known as the Great Plains Wolf. Many of the wolves in this population are radio collared, and Minnesota's North Central Forest Experiment Station has been a key element in the study of predator/prey relationships between grey wolves and white tailed deer in this area, for over twenty five years. Now, as I neared my final destination in a town called Ely, I was about to spend ten days of my life, studying and radio tracking those same wolves and deer. Starting my own wolf research back in 1989, I now realised a dream was becoming reality. I had, indeed, arrived in wolf country.

Welcome to Ely

I was a stranger in a strange land, but from the time I arrived I had begun to feel welcome. I was to spend the next ten days based in and around Ely, as a member of an Earthwatch team, monitoring the movements of nearby wolves through radio tracking.

Ely itself is situated in, what you might call, 'the heart' of the Superior National Forest. This is a huge expanse of boreal woodland covering over two million acres, from Northern Minnesota to the Southern Ontario border. Ely is also the home to 'the International Wolf Center', a true 'Mecca' for any wolf devotee. By American standards, Ely is a small town, with a population averaging 4,000 plus, but Ely is also a wolf town. Everyone knows of the wolves in the outlying area, and indeed, many of the town's inhabitants depend on the wolves for their livelihoods, especially when it comes to tourism. Sightings of wolves are still not so common though in the outlying area, because of the wolves' elusiveness, and a new sighting always brings about exciting conversation with the locals. As I sat having my breakfast in the 'Chocolate Moose' cafe on that first sunny morning, I listened to the cook tell his regular customers of how, on the previous night, he had caught three wolf pups in his vehicle headlights. He had then turned off the engine, and he and his two children had been enthralled by the pups as they played happily for over half an hour. I smiled warmly. Tales like this were just so special in a town like Ely. "At last" I thought..... "I really have arrived, wild wolves were not thousands of miles away, but all around me". My spine tingled with excitement as a car pulled up next to the cafe and a gentleman came in to greet me. "Hi Ian, I'm Dan Groebner and I believe you've come looking for wolves". "That's right" was my reply. "Well you've certainly come to the right place" said Dan. "Indeed I have" I thought.

The wolf in Minnesota

By the year 1965, the only viable wolf populations left alive in the Lower 48 States of the United States of America were in the extreme north-eastern sector of the Superior National Forest in Northern Minnesota. All the other wolf populations across this vast expanse had been totally eradicated, or reduced to non-breeding populations of individual animals. The U.S. Biological Service and other government bodies had done their job extremely well. Over a period of one hundred years or more, hundreds of thousands of wolves had been shot, poisoned, bated and trapped, all in the name of predator control.

However, the situation regarding wolf control in Minnesota had been slightly different up until 1965. Because there were no ranchers and herdsmen, and because of the huge inaccessible expanse of boreal forest expanding from Northern Minnesota into Canada, grey wolves still had a chance of survival, against human persecution. By this time the total population of wolves in this area probably ranged from between 400 to 800 wolves, but even these animals were still being partially controlled under a 'predator control programme' between 1969 and 1974. However, with the passing of the Endangered Species Act (ESA) in 1973, all this changed from 1974 onwards, as the grey wolf became listed as an endangered species. After the passing of the ESA, a joint team of federal and state personnel and biologists both working for the U.S. Fish and Wildlife Service and the Minnesota Department of Natural Resources devised a wolf recovery programme, initially entitled the 'Eastern Timber Wolf Recovery Programme'. The year was 1975 and, as the years moved on, the population of grey wolves continued to expand, now under legal protection and the control of 'specialist' managers who had the wolves' future survival in their minds and indeed, in their hearts. The plan itself became a huge success, and although small numbers of wolves were removed due to livestock predation, the actual population in Minnesota grew and was downlisted to 'threatened' status. By 1989, thanks to protection and specialised management, the wolf population in Minnesota had expanded to between 1,550 and 1,750 animals. This population had now exceeded the goal of the recovery programme which had been set at between 1,251 and 1,400 animals. Now, various wolf packs were expanding and dispersing rapidly into specially managed zones from north to central Minnesota. This recovery programme for grey wolves has been a remarkable success, with approximately 2,200 wolves now inhabiting the State. Neighbouring Wisconsin and Michigan have also seen rapid recovery of their wolf populations through naturally dispersing wolves moving into these areas, many of which are from Minnesota.

The recovery of grey wolves in the states of Minnesota, Wisconsin and Michigan's Upper Peninsula, collectively known as the Great Lakes area, has certainly been a triumph for human ingenuity and understanding. That triumph has been achieved by many professional people, working with volunteers for many years, to help collect and then pass on to the general public, data and information that has helped lead to recovery and understanding of this magnificent wild dog. As I joined Dan Groebner and my other nine colleagues in the land of the grey wolf, in June 1992, I knew I was playing my part in that recovery, a part that still lives on in my heart and my imagination to this very day.

Nightime thoughts

The next three days were a mixture of transferring to the tracking project headquarters, learning to map read and utilise radio telemetry information from radio collars, and of course, making new friends, all of whom were American. Six volunteers had joined me and I was intrigued by their backgrounds, which were as diverse as a traindriver, insurance salesman, telephone engineer and company receptionist. Over our three days of basic training, we had all become good friends, and were now eager to get out in the vans and start our first twenty-four hour tracking session.

On the night before the tracking started, I sat outside on the veranda of our log cabin. My thoughts were jumping back and to as I waited for the dawn to arrive. Fireflies had started to weave their magical dance and nighthawks swooped low overhead. I smiled and looked up at the Minnesota night sky. The stars, thousands of them, were gleaming like diamonds, and a half moon shone brightly overhead. As Mother Nature enveloped my senses, I started to wonder what adventures lay ahead. It was just then that I heard a faint, low howl away in the trees. Several of my friends heard it also. The wolves were calling to us, or so it seemed. I picked myself up and moved to my bedroom. It was hard to sleep that night.

"Let's go find our wolf"

The phone call came in at 6.30 the next morning. Dan Groebner had answered, and gained the information he had been waiting for. A research biologist had received information from Dr. Mike Nelson that wolf number 369 had now moved back into our area, and we were to start tracking his movements, as of now.

Dr. Mike Nelson is the wildlife research biologist who started the wolf and white tailed deer ecological studies in Northern Minnesota over twenty years ago, along with Dr. L. David Mech. Dr. Nelson's white and red 'spotter' aircraft is well-known in the skies over Minnesota's wolf range, and the information regarding wolf 369 had come from that same aircraft, flying at 2,000 feet approximately 10 miles away from us. The number given to the wolf we were to track came from a metal tag which had been attached to one of the wolf's ears several months before, by Dr. Mech and his research technicians. Number 369 had recently dispersed into Canada, but now he was back in his old territory. Now it was time to put our training to the test.

It was 7.00 a.m., the sun was shining and I had never felt so excited. Hardly any of us had slept through the night, and now, we were all outside with Dan Groebner and fellow instructor and researcher John 'Denfinder' Brugginck. Our two tracking vans, with their large aerials looking like giant antennae, were all cleaned up and ready to go. I had volunteered for two shifts, the afternoon and the nightime. The first teams were now in the two vans, collectively known as Gamma and Omega. It was a two or three person job to do the tracking and everyone had different shifts. At 8.00 a.m. the vans started to roll. I waved my friends off as we all clapped and cheered. "Let's go find our wolf" shouted John Brugginck who was driving Gamma. The vans moved past me and out onto the road, then off towards the area where wolf 369 was last sighted. Within an hour, they would be well off any major road, and deep within the forest. My adrenaline started to pump rapidly and I ran back into the project cabin where Dan was monitoring the radio communication between ourselves and our colleagues in the vans.

"Hello Gamma and Omega, this is base, do you have a copy yet, over?". You could sense the excitement and anticipation in Dan's voice as the radio crackled with static. "Negative base, we're still checking, over". Fifteen minutes later, the action commenced. "Base, this is Gamma, we have a moderate to strong signal, 369 is in range and locked on........it's looking good".
"Roger that Gamma......" said Dan "stay as close as you can and keep the readouts coming through".

The vans had indeed gained a strong signal. The collared wolf was down by the aptly named 'Wolf Lake' and his movements were now being monitored very carefully. Now my adventures were really beginning, but little did I know what lay ahead, as the evening started to draw in.

On the hunt

The blip on the radio monitor was strong. It was 6.30 p.m. and I had been tracking since 2.00 p.m. with Dan Groebner in Gamma. We were taking readings of the wolves' movements every fifteen minutes by swinging the large aerial around on a 360 degree angle. The signal from 369's radio collar sounded strongest at a certain compass bearing and we would then radio our co-ordinates to our companions in Omega, who would then verify their co-ordinates with us. We would match our settings on the map of the area which was installed in the vans, and pinpoint where our wolf was. For most of the afternoon, wolf 369 had been sleeping, and our vans had been in the same positions for several hours. Wolf 369 belonged to the so-called 'Armstrong Pack' which had a den just over a mile away from where we were tracking. We were right in the thick of the forest, and Dan had worked hard and carefully to get Gamma into the position required. It was a delight to have no traffic near us at all. Now and again the odd white tailed deer or snowshoe hare would move leisurely past the van, with ourselves keeping very quiet and still. Skunks were also common, as were beavers, coyotes and, on rare occasions, black bears. It was a beautiful and moving experience to know that I was right in the heart of true wilderness. As the hours ticked by and the darkness crept in, the Aurora Borealis, or Northern Lights, lit up the sky, caressing the stars in a glowing curtain of shimmering colour. I stared in amazement at the splendour of it all.

As the daylight departed, we knew that the real excitement of radio tracking wolves was about to begin, for the nightime brings 'the hunt'.

* * * * * * * * * * * * * * *

It was 9.15 p.m.

"Gamma this is Omega, we have active movement, suggest 369 is on the hunt, over".
Dan replied "yes copy that Omega, he's moving fast, pace yourselves and don't lose the signal, we're moving out also".

So this was it, Dan had told me to expect a rough ride throughout the night, and it was now beginning. Wolf 369 had moved rapidly northwards, and 9.30 p.m. found us crashing through mud trails and logging routes to try to keep up with the signal. "Hold on Ian" was all Dan could say as our flasks of coffee smashed onto the floor of the van. "I know exactly where he's going" shouted Dan, amidst the cacophony of a straining engine and crunching tyres. "He's heading for a herd of white tails approximately five miles away, I figured he'd do this". Dan hauled the heavy steering wheel around, then yelled alarmingly "a bull moose!".

I strained my eyes into the beam of the headlights. My adrenaline was pumping rapidly through my body.......yes......sure enough.....a huge bull moose had moved right in front of the oncoming van. Dan slammed the brakes on, and threw the van into reverse. "We daren't go near that guy.....he'd ram us for sure......always stay well clear of moose". Dan was smiling, I could tell he was thoroughly enjoying the excitement of radio tracking a grey wolf at night, and so was I. "We're losing the signal" I shouted....."Hello Gamma, this is Omega.....are you guys OK?"
Our colleagues had obviously become concerned at our lack of radio contact. I replied "hello Omega, yes, we're fine, just a few problems with a moose......that's all". Dan laughed, and so did I.

Thirty-five minutes later we found our new position, thanks to the help of our colleagues in Omega. Wolf 369 moved on the hunt throughout the night, and it was fascinating watching his journey and hunting technique taking shape on the map, as we plotted his movements from that wonderful tell-tale signal emanating from the wolf's radio collar. The night itself consisted of several 'chase scenarios' and we moved position at least nine times over a period of five hours. By 6.00 a.m. the next morning, all was calm. We were down by a small lake, and mist hung all around us. The windows of the van were wound down, and large dragonflies flew in and out, sometimes landing on our shoulders. There were only another two hours of tracking left for the session, and 369 was now asleep again. He had certainly given us all a night to remember, and would do so again as the week progressed.

By 8.00 a.m. the sun was shining and the forest was alive. "O.K. guys, time to wind up for the moment" said Dan over the radio. Dan smiled at me cheerfully "not bad for your first night wolf tracking Ian......not bad at all".

Dan slid the large side door of the van open, and we both got out and took a long luxurious stretch "ahh" I said...... "it's going to be a beautiful day............".

CHAPTER 3

Savage Freedom
Wolves on the hunt

"To understand the true persistence, determination and ability of a moose-hunting wolf, one needs to watch a single, hungry wolf at work".

Bob Hayes Canadian Wildlife Biologist
'Hunters of Moose' (International Wolf Magazine - Summer 1993)

'A world of savage freedom'

The moose stood its ground. So too, did the wolves. Five of them. The ice encrusted snow crunched under the huge hooves of the intended prey. The bull moose looked around, surveying its antagonists. Its thick neck muscles thrust downward as the moose lowered its huge antlers towards the three larger wolves. If it stood its ground, the wolves would have to think carefully before attacking, for this bull moose was in its prime. It showed no weakness, no fear, no apprehension. The moose bellowed defiance through its huge nostrils. The rear wolves attacked...............
A massive rear hoof flashed out, connecting with a canid rib cage. A howl of fear and pain echoed across the forest as the other wolf rolled across the ground under the lethal weaponry of the moose. The alpha wolf barked at its packmates. A warning sign. The prey had become the predator. Today, there would be no hunt. Today, the moose would go its own way, and the wolves would go hungry. For the world of the grey wolf on the hunt can, at times, be intensely savage, and yet, this is a world of freedom that we humans will never know. It is a world of savage freedom.

What really are the facts behind this one aspect of wolf behaviour, ecology and society that still, in many ways, eludes us all...................the hunt. My intention in this chapter, is to take you into the world of the grey wolf on the hunt in America and Canada. Through first class biological references we will enter the realm of predator/prey dynamics between four of the five sub-species of North American grey wolf and their various prey.

However, this chapter will also give a meaningful insight into the hunting instinct and capabilities of all grey wolves, whether they inhabit North America or anywhere else. Sections on the tools of the trade, prey search imagery and the pack itself are all reminiscent of wolf behaviour, across the Northern Hemisphere.

So now, let us enter the world of the grey wolf on the hunt, and discover the truth about this fascinating aspect of wolf society.

'The tools of the trade'

"Out of every ten attempts by a hungry wolf pack on the hunt, only one or two will be successful".

Professor Rolf Peterson (Isle Royale Wolf Studies, Michigan)

The grey wolf is a top 'apex' predator in the Northern Hemisphere and, in reality, only has three main enemies..........these are: other wolves, bears and humans. Because of this fact, the wolf is classed as a carnivore at the top of the food chain, and yet, Mother Nature still keeps most of the wolf's prey species one step ahead of actual predation. So the wolf must work hard to secure food and, ultimately..........survival.

In the wolf's world of survival of the fittest, the tools of the trade are in constant readiness and must be honed to perfection. Let us first take a look at these 'tools' of predation and, what better way to start, than with the wolf's teeth and jaws, the 'cutting edge' of savage freedom.

The wolf has forty-two teeth. These consist of the frontal canines and incisors, the pre-molars, carnassials and molars. The four large, pointed canines and incisors are the main teeth which cause grievous bodily harm to the intended prey. The premolars and molars are used for grinding up bone and tissue, whilst the carnassials help with the sheering of meat from the bone. The actual dental formula for the teeth is as follows:

        Upper teeth (each side)   3-1-4-2
        Lower teeth (each side)   3-1-4-3
    incisors-canines-premolars-molars (including carnassials)

As well as teeth, the wolf's jaws play a vital role in carnivory, and can exert a bite pressure of over 1,500 lbs per square inch. That is twice the pressure of a German Shepherd Dog's bite. Also, whilst on the hunt, the wolf must utilise all its predatory senses. Before and during the hunt, body language also plays an essential role in communication, especially for the pack's use of strategy and tactics. In fact, the wolf's basic forms of communication are far more advanced than ours and their co-operative hunting strategies show this fact to the full.

The Russians have a saying that "the wolf is fed by its feet", and in a way, this is true. The wolf's stamina is quite amazing, as they can 'lope' constantly at around six miles per hour for many miles at a time and, on occasions, can cut in to bursts of speed, measuring up to forty miles per hour whilst chasing prey. Fine examples of this type of behaviour can be seen in Alaska, where packs of wolves follow migrating caribou, sometimes for hundreds of miles. Wolves are also excellent swimmers, and have no fear of water when it comes to hunting down prey. Although a large majority of the wolf's prey species will seek safety in water, on occasions, wolves will enter water and swim after their prey, actually attacking and killing whilst in this aquatic environment. This type of hunting technique has been fully documented in the Superior National Forest in Minnesota, Montana and Wood Buffalo National Park in Alberta.

Whilst looking at the 'tools of the trade' for the hunt itself, we must take a look at the wolf's use of 'prey search imagery', for this is the main key element for the selection, and hunting down of the wolfs' prey species. The concept of prey search imagery was first developed by Lukas Tinbergen who studied the prey selection of various birds. Through his theory, he tried to explain why certain birds tended to choose their kills from more abundant prey than with other prey that appeared in low densities in the same predation area. This 'search image' was thought to be a type of learned visual filter which allowed predation on just one or a few prey species to the point where other prey were ignored. This concept was then taken one step further by Dr. Steven H. Fritts who then theorised that something like this was happening in certain areas of wolf habitat in the United States, especially with low livestock loss. This meant that the wolf's imaging system basically recognised main prey species such as white tailed deer and elk, but at times did not relate to a common food source such as livestock. Dr. Fritts's studies in Minnesota showed that in certain areas of wolf/farmland habitat, livestock loss was minimal. This was then attributed to the grey wolf's use of prey search imagery being a strong deterrent to depredation on domestic livestock. In other words, at most times, the grey wolf has a strong in-built image of its main prey species.

It is also important to note that on most successful hunts, wolves will kill old, sick, weak, infirm and very young examples of their prey thus keeping and maintaining a healthy balance of prey in any given area. The keen eyesight of wolves helps to correctly identify any prey animal which shows signs of vulnerability such as old age, damaged joints, arthritis or sickness. During this part of the hunt, commonly called 'the confrontation stage', wolves will test their prey. If at any given time the prey shows no sign of weakness, sickness etc., they will probably move on to another herd (although, on occasions, wolves will kill healthy prey if all else fails).

And finally, we must also take a look at another important 'tool of the trade' for the hunt, and that is the wolf pack itself................the wolf pack is, quite simply, a hunting and killing machine. Every wolf within the pack knows its place, and the main senior members of the pack, especially the alphas and betas, will utilise an amazing degree of intelligence concerning the use of strategy and tactics for the hunt itself. Once the pack starts the hunt, an incredible amount of co-operation will, hopefully, bring success. However, life on the hunt for the wolf is particularly tough, as Mother Nature has given the wolf's prey species various defence mechanisms to combat wolf predation. So now, let us move on, to the hunt itself, and to the various prey species of wolves in North America..............

"**Let us prey**"

"*Then, I couldn't believe it..............fifty yards behind the musk oxen, far away on the hillside, ran another wolf. She was after the third calf!!*"

L. David Mech   'The Great Musk Ox Hunt'
(Defenders Magazine - Sept/Oct 1988)

In the vastness of North America, the grey wolf has many prey species to hunt. These can vary considerably, from the mightiest of bison to the smallest of rodents. Also, at times, wolves will show omnivorous tendencies, with the eating of vegetable and fruit matter. However, in general, the wolf is a carnivore.......it is a meat eater........and it must hunt down other species to survive. So now, let us enter the world of 'predator - prey dynamics' between the grey wolf and its major prey species in North America.

According to world renowned wildlife research biologist Dr. L. David Mech, there are six stages to the hunt itself, these are:-

    1.   The locating of the prey
    2.   The stalking of the prey
    3.   The confrontation

4.   The attack (causing the panic factor)
5.   The chase
6.   The kill

With these six aspects in mind, we will move on to our first scenario and take an in-depth look into the world of predation of Canis lupus arctos (the arctic wolf) on Ellesmere Island:

i)   Wolves and musk oxen (Ovibus moschatus)

Musk oxen are certainly a major prey opponent for the arctic wolf. These huge animals can weigh anything between 700 lbs to 900 lbs. The musk oxen is a herding animal and generally seeks safety in numbers, utilising two types of defence mechanism, the first of which is 'the single line defence'. This line faces full-frontal to the opposing wolf pack and consists of the bulls at the front, ready to defend the cows and calves at the rear with their deadly hooves and horns. The bulls will continuously shift the line to match the movements of the wolves. However, if the pack then starts to circle the herd, the bulls will 'close ranks' and form a circular base with the cows and the calves in the middle. Once again the bulls will stand defiant against the wolves and will charge any wolf that starts getting too close. At this time, the predating wolf pack will then start utilising search image to seek out suitable targets within the herd. Scientific studies in the High Arctic have shown that musk ox over the age of ten years are highly susceptible to predation once their defences have been broken by the pack. Also calves are a prime target once the herd are thrown into what, I call, a 'confusion mode' where the wolves have moved into stages four and five of the hunt. Once panicked by the pack, the herd will flee and the wolves can then try to pick off their intended prey. However, 'stand offs' between predators and the prey, as in many other predation scenarios, can last for hours. In this situation, the wolves may well move on if the defence mechanism is strong and cannot be broken.

ii)   Wolves and arctic hare (Lepus arcticus)

Another staple prey species for the wolves of the High Arctic. Arctic hares can weigh up to twelve pounds and live in large groups. Their main defence mechanisms against wolf predation are speed, agility and constant observation of their outlying area. Young arctic hares also blend in perfectly to the terrain staying in a 'frozen' state until the predator has passed. However, once the hunt begins, arctic wolves will generally concentrate on one individual from the group, and give chase using a 'pincer' type movement, as one wolf will

chase the hare into the path of another. Once again, the intelligence of wolves comes into play with the use of strategy and tactics. Arctic wolves will also use their keen sense of smell to locate young arctic hares lying still.

iii) Wolves and bison   (Bison bison)

Weighing in at anything from 400 lbs to over two tons, the North American bison is by far the most formidable prey species for the grey wolf and is also North America's largest land mammal. The main area of research for the study of predator prey relationships between these two distinct species is Wood Buffalo National Park in Northern Alberta where the grey wolf subspecies 'occidentalis' inhabits. Although there are also bison in Yellowstone National Park, biologists agree that these bison are relatively safe from wolf predation because of the large amount of deer and elk in the Yellowstone Park area. The North American bison is a truly dynamic animal. Bulls typically weigh anything from 800 lbs to over 2000 lbs. Cows vary in weight from 400 lbs to over 1000 lbs and calves weigh in at around 30 to 40 lbs at birth. Like the musk ox, the bison is a herding animal and utilises strength in numbers. The bison's main armoury are its hooves and horns with their main defence mechanism called the 'stampede defence'. This involves the selected herd of bison stampeding as the pack gives chase. The bulls will then try to confuse the wolves by trying to cover the cows and calves whilst stampeding, keeping the calves well into the middle of the herd itself. When bison stampede in this way, the predating wolves will give chase from both sides and from the rear, trying to separate out the chosen prey animal. The wolves will then attack the rear and flanks of the bison then leave the animal, with quite possibly, a large amount of blood loss, this will then lead to the death of the bison whilst the wolves wait at a safe distance. Bison are also excellent swimmers and seek safety on many occasions by swimming away from their antagonists. However, as I mentioned previously, wolves will, on occasions, follow the prey into the water, still continuing the attack. Again, weakness in the herd itself plays a major part in prey selection, but wolves will prey on calves quite intensely during Summer months. Mature, healthy bison are also preyed upon by wolves, but this is not so common. One fascinating aspect of wolf predation on bison is the use of 'multiple kills'. This is a type of hunting technique used by large packs, where the wolves will drive the bison into 'chosen' areas, then another part of the pack will lie in wait to 'pick off' the advancing bison, especially the calves. This chosen type of predation seems to work well in certain locations of Wood Buffalo National Park.

iv) Wolves and caribou (Rangifer tarandus)

The caribou is a highly migratory animal, sometimes travelling hundreds of miles with wolf packs closely following. The caribou inhabits vast areas of Canada and of Alaska, and is a main food source for the arctos, occidentalis and some nubilus sub-species of grey wolf in North America. Groups of migrating caribou can vary from several hundred to several thousand, and it is this colossal amount of numbers that gives a main defence mechanism against wolf predation. Males usually weigh between 200 to 400 pounds and females between 125 to 250 pounds, but both sexes may grow much larger. Caribou calves will vary in weight from between 12 to 100 lbs depending on age. Because of the close proximity of wolves with the migrating herds, it seems to be a fact that, most of the time, the caribou show no anxiety in relation to nearby wolf activity. However, when the hunt begins the animals within a few hundred yards of the predating wolves will show concern. There is also a major distinction between the caribou and the grey wolf's other major prey species in North America. This is due to stage two of the hunt, 'the stalking of the prey'. Because of a great deal of 'open' habitat in certain areas of predation, wolves will stalk caribou far more than with other prey species. Like a lion stalking a gazelle, a grey wolf (or wolves) will stalk the caribou as low to the ground as possible, and then, at the last instant, will give chase from a major vantage point, just a few yards from the intended prey animal itself. This is actually classed as an 'ambush' technique. Wolves will also hunt down caribou utilising two other types of predation which are similar to hunting other prey species such as bison and musk ox. These are:

i) relay running, in which wolves will tire the prey animal by constantly moving it into the path of other pack members who will then continue the chase.

ii) the 'full chase'of the herd itself attacking from the rear and side, then singling out the weaker animals. However, in general, these two hunting techniques do not have the same success rate as the 'ambushing' technique on caribou.

Once again, calf predation takes its toll during the spring and summer, and especially during the first week of birth, when the female will move away from the herd to give birth. The newborn calves are then completely helpless for a week or so, and predation rates can be high in certain areas. However, on a yearly basis, a variation of both adults and calves are taken by predating wolves and also by brown bears.

Another interesting aspect of the predator/prey relationship between the wolf and the caribou is the rate of attack by lone wolves on caribou herds. Because of the increased density of caribou herds compared to the wolf's other prey species, lone wolves have a far higher rate of success on the hunt with caribou.

v) **Wolves and wild sheep**
    Dall sheep   **(Ovis dalli)**
    Bighorn sheep   **(Ovis canadensis)**

Inhabiting Alaska and the Northern Canadian Rockies (dall sheep), and the Central Rocky Mountains (bighorn sheep), these prey species utilise their precipitous terrain for defence against wolf predation from the occidentalis and nubilus subspecies of grey wolf. On average, bighorn rams weigh up to 320 lbs and ewes up to 175 lbs. Dall rams weigh 200 lbs and ewes slightly less. These species of wild sheep are extremely agile on their terrain and will usually try to move further up a mountainside if confronted by wolves. However, as observed on certain occasions, wolves have learned to cut wild sheep off from the top, and circle the one that has been singled out. This use of strategy is generally successful for a large pack of six plus wolves or more. Also, if wild sheep are caught out in the 'open' areas, then predating wolves do have a good chance of running down and tiring them. In an amazing study in the Yukon in 1987, wildlife biologist Ronald S. Sumanik actually recorded wolf predation on dall sheep for several months. His study showed that wolves utilised strategy and tactics for predation in five different ways. These were:-

    i)   predation on open slopes
    ii)   predation in shallow ravines
    iii)   surprise 'ambush' predation on creek or river ice
    iv)   forced defensive running of sheep straight off a cliff face
    v)   downhill open slope pursuit.

However, in general, wild sheep do seem to have a high success rate of escaping predation by wolves.

vi) **Wolves and elk (Cervus elaphus)**

More commonly called the Wapiti in Canada, the elk is a staple prey element for the occidentalis and nubilus sub-species of grey wolf. Males, on average, weigh approximately 700 lbs and females 500 lbs. Again, the elk is a herding

animal and seeks safety in numbers. Other defence mechanisms against predation are alertness, speed and sometimes on sheer size and strength. Bull elk will also stand their ground against wolves. Wolves hunt elk very much the same as deer, utilising the 'panic factor' and the 'chase' as main tools of predation. However, biologists seem to agree, that elk can be a difficult target, unless they are very young or are affected by snow conditions, old age or illness. In the hunting and killing of elk, wolves will usually attack the rear and the side of the animal, but the nose and the throat are also useful target areas.

vii) Wolves and deer
White-tailed deer (Odocoileus virginianus)
Mule deer (Odocoileus hemionus)

Certainly the most important prey species for the wolves of North America are the two main members of the deer family. These are the white-tailed deer and the mule deer. So widespread are these two species, that they form the most abundant prey base for the occidentalis, nubilus and lycaon sub-species of grey wolf. Deer are actually the smallest of the ungulate prey for wolves in North America. Adult white-tailed bucks can weigh anything from 150 lbs to 250 lbs, where does average between 90 to 120 lbs. Mule deer weigh in approximately one fifth more than white-tails. Although there is not a great deal known of the interaction between wolves and mule deer, the predator/prey relationship between wolves and white-tailed deer has been intensely studied, especially, in the Superior National Forest in Northern Minnesota. In general, deer have three main defence mechanisms against wolf predation, these are speed, agility and their incredibly acute sense of hearing. It is this latter sense that keeps all deer constantly on the alert for predators. Certainly, the largest amount of deer that fall to wolf predation are old or sick animals, or very young fawns. Deer will also gather in large herds in wintertime, giving the defence strategy of strength in numbers.

viii) Wolves and moose (Alces alces)

The moose, just like the bison, must be classed as another formidable opponent for grey wolves, as they are certainly the largest of their antlered prey. The bull moose itself is an imposing, incredible animal and can weigh up to 1,250 lbs. Cows can weigh up to 850 lbs and calves from 25 to 300 lbs depending on age. Wolves that have experience of many hunts know that healthy moose can cause them grave injury, especially if moose stand off and face their canid opponents. Indeed, one Canadian wildlife biologist has likened the moose's hooves to 'sledgehammers', but if the 'target' moose starts to run, then predating wolves will give chase, knowing this time that their prey may be vulnerable. Calves are also heavily hunted by wolves, but cow moose will defend their offspring vigorously throughout their first year of life.

The most intensive study of wolf-moose interaction has to be the legendary Isle Royale ecological studies, first started by Professor Durward L. Allen and L. David Mech back in 1957/58. These studies were then fully taken over by Professor Rolf O. Peterson from 1970 onwards. This unique study has given light to many details between predating wolves and defensive moose dynamics. In general, moose are disadvantaged if they start to run from any hunting wolf pack. If the moose tries this defensive strategy, it gives the wolves a perfect chance to attack the rear and the flanks. Then, as the moose weakens, the wolves will move to attack the nose and throat. Once fully weakened and in a state of shock, the wolves will then leave the moose, possibly for several hours. During this time, the moose will then die, through shock and loss of blood. Certainly the best defensive strategy for moose, against a predating wolf or wolf pack, is to stand its ground, ready for a fight. If this happens, most of the time the moose will be left alone and the wolves will move on. Moose will also seek safety from wolves on the hunt by entering water, but sometimes, this may not prove fully effective.

We have now looked at several major aspects of predator/prey interaction between grey wolves and their main prey species in North America. Of course, there is other prey for wolves in this area, such as beaver and snowshoe hare, but we must now move into two very different areas of savage freedom. Let us now take a look at the rules of territoriality, where wolves will, at times, hunt other wolves.

"In defence of the realm................Stay out and keep out"

"Grey wolves face the most risk of violent death.........from other wolves"

Jay Hutchinson   Strife among wolves: When aliens meet (International Wolf - Winter 1994)

When we look at the world of the grey wolf on the hunt, we must cover the use of territoriality. When wolves claim territory, they claim a piece of land which is their hunting ground. This area is mapped out using scent marking from the alpha wolves' anal and paw glands. The message of scenting is then picked up by other animals coming close to this territory. When wolves lay out territorial boundaries, they send a message which is brutally simple "stay out and keep out.........if you do not.......we will hunt you down, attack.......and, most probably, kill you!".

There have been many occasions where wolf research biologists have witnessed the hunting down, and killing, of trespassing wolves in another packs' territory. Other carnivores will not be tolerated either. One of the most striking examples of wolves repelling and then hunting down trespassers in their territory, comes from a major study of the interaction of wolves and bears at a wolf den site in the Northern Yukon, by Canadian biologists Bob Hayes and David Mossop. In this absorbing and often startling report, the two biologists observed a sow brown bear and her cubs, estimated at between 2-3 years of age, digging for wolf pups at a den. They were then surrounded by seven members of the pack, who attacked the three bears. The wolves then chased and attacked the bears again, pursuing them for a total of approximately 300 metres. One of the wolves was seen taking a bite at one of the bear cubs, then holding on for at least 10 metres before being thrown off. This type of study has shown that, although wolves and bears do not selectively hunt each other, at certain times they will go 'all out' to kill each other, especially given the rules of territoriality.

There are two main exceptions to the rules of territoriality, however, whereby unfamiliar wolves will be allowed into another pack's territory and it is important to cover these aspects of wolf behaviour in this section. Many wolf research biologists across the Northern Hemisphere have proved that, in general, wild wolves do not like to inbreed as they need to strengthen their gene pool during the mating season, to help keep their offspring as strong and as healthy as possible. Therefore, dispersing wolves that leave their pack looking for a mate may well bring an unfamiliar wolf back to their natal pack if they cannot find any free territorial area to breed. This wolf will then, most probably, become a breeding mate in the future, thus helping to strengthen and stabilise the pack's gene pool.

Lone unfamiliar wolves may also be allowed to join a pack on occasions, in an area of high prey density and low wolf pack density. This is an area free from major pack strife over hunting territory due to the large amount of prey available. This wolf will be allowed to join the pack to assist in daily routine matters such as caring for the pups and joining the hunt. Again, this wolf will probably help with the breeding process over future mating seasons.

With the exception of these two important aspects of wolf behaviour, when looking at the rules of territoriality, it is a fact that wolves will defend their territory to the death, especially in defence of their pups.

'The livestock factor'

"The problem of wolves killing livestock presents the greatest challenge of acceptance of the wolf"

Steven H. Fritts Livestock depredation - The downside of wolf recovery (International Wolf - Spring 1993)

In the annals of wolf research, it is now a well-known fact that grey wolves do, on occasions, kill livestock. Although wolves do not 'hunt' livestock in the way I have previously described in this chapter, it is still very important to look into 'the livestock factor' as such, as it is an aspect of the wolf's survival that we cannot ignore.

So, what are the facts behind wolf depredation on livestock? As far as I am concerned, I have noted through my years of wolf research, that there are three main reasons why wolves will kill livestock, these are as follows:

i) Livestock depredation in areas of low prey density
   This is an area where the wolf's normal prey species will be low in numbers. The wolf packs, or lone wolves, may be struggling to survive because of this, and so, given the opportunistic predatory behaviour of the wolf, depredation on livestock may well occur.

ii) Rogue wolf syndrome
   This is actually my own term for wolves that seem to throw away the rules of 'prey search imagery' and start preying on livestock as an 'easy' food source. 'Rogue' wolves will probably keep returning to kill livestock, and as such, will probably end up being trapped or shot by an Animal Damage Control biologist, under the rules of wolf management.

iii) Sick wolves and livestock
   Wolves can suffer from a great many maladies. From canine parvovirus, sarcoptic mange to worm infections. Any maladies that sap a wolf's strength will leave that wolf at a disadvantage when it comes to the hunt itself. Therefore, sick or weak wolves may kill an easy food source such as livestock.

These three key points regarding wolves and livestock add another piece to the puzzle of understanding the survival instinct of wolves, and the hunt itself. The livestock factor also presents human beings with a major challenge regarding our relationship with the wolf, especially in the case of reintroducing wolves into the wild. Though controversial, a viable management plan for depredation on livestock must be in place for wolf recovery to occur. In the United States wolf management plans regarding the loss of livestock use a variety of approaches depending on the cultural values of the region where the depredations occur. These include leaving the wolves alone, capturing and then moving them to another location or euthanising the depredating wolves. A private conservation group in the United States called 'Defenders of Wildlife' also helps to pay compensation to farmers who have lost their livestock to wolves. This economic incentive also creates tolerance within the farming community and so helps promote wolf survival.

In the world of the grey wolf on the hunt, there will always be blood on the snow and carcasses on the ground, but that is their way. It is a way of survival and a way that we must all come to understand. Once this way is accepted by our kind, we should rejoice in the knowledge that there is still, in wilderness, a realm of savage freedom.

CHAPTER 4

A Howling Over Russia
Tracking Wolves The Hard Way

The bread was stale, the cheese was strong and the tea tasted of aniseed. The tree trunk, with my food upon it, was covered in ants. Mosquitoes were everywhere and, away in the distance, woodpeckers tapped out their concentrated rhythms on the trunks of their Scots pine homes.

It was 10.15 p.m. on Monday the 9th of August 1993, and as I sat eating my supper, I could not help but think of how good it felt to be back in wolf country. My two Russian friends, Vitali Kochetkov and his assistant Sasha, had only met me two hours before, and yet, as Vitali thumbed his way through his Russian/English dictionary, I felt I had known these two fine gentlemen for years. Indeed, we were to become the very best of friends as they both took me on a splendid weeklong journey, into the secret world of the Russian wolf.

Little did I know at that precise moment in time that through the night ahead we would be following a wolf pack through forest and fields as the adult wolves moved their puppies to a new home area, called a rendezvous site, and that by the end of the night I would be utterly exhausted. The effort, however, would prove to be worthwhile as, unknown to me at that time, I would be just thirty feet away from a wolf pack on the move. The world's largest forest was about to welcome me with a grand adventure.

As I finished my supper and gathered up my belongings ready for the first night of wolf tracking, my mind raced back to that great moment in human history, when the Berlin Wall came crashing down and East finally met with West. Now, here I was, in the middle of Russia, working with Russian wolf research biologists. The 'Iron Curtain' no longer existed. I walked away with Vitali and Sasha, as two Russian MIG 29 Jet Fighters screamed low overhead. Five years ago this whole scenario would have been unthinkable. Now, the old Soviet Union was no more, and it was time to take my wolf studies one step further. "Welcome to Russia" I thought.

A Russian wonderland

It is the world's largest forest. It stretches across Northern Russia, from the Finnish Border to the Pacific. It crosses through three different time zones, and is twice the size of the Amazon Rainforest. It is the Taiga and, on Sunday the 8th August 1993, myself, my niece and ten other fellow adventurers found ourselves in the north west area of this immense forest, over 400 miles from Moscow. We had arrived at the I.U.C.N. Biosphere Research Reserve in the old Soviet district of Nelidovo, and quite literally miles away from any main core of humanity. Although my niece and fellow colleagues had arrived at Nelidovo for an intense week's study of all aspects of the flora and fauna of the Taiga, I was there for one specific reason only, namely the in-depth study of the Taiga's wolves.

The Taiga itself really is a Russian Wonderland, utilising thousands of different species of plants and wildlife. This huge boreal expanse of forest is also home to the last great remnants of wolf pack activity in the whole of Europe. Although there has never been a fully completed census on the total wolf population of Russia, the Taiga most probably holds between 40,000 to 50,000 wolves. It is here, in the huge mass of woodland, stretching up to 1,500 miles in width across Russia, that the wolf reigns supreme as the top predator, preying on the Taiga's abundance of European elk and reindeer (more commonly known as moose and caribou in North America). The fauna of the Taiga is incredibly diverse, ranging from the predators at the top of the food chain such as wolves, brown bears and lynx, down to the mainstay prey animals such as elk, reindeer and beaver. One will also find such wildlife here as wolverines, pine martens, red and arctic foxes and a profusion of birds and rodents. The flora of the Taiga consists mainly of pines, larches, spruces and firs, providing a massive canopy of oxygen-radiating, life-giving, boreal forest and a safe haven for thousands of wolves. The wolves of the Taiga, as in all their different habitats, play an important role in the ecology of the Russian woodlands, keeping herds of prey animals at consistently strong and healthy levels, and thus ensuring a healthy wildlife balance in this most beautiful of forests.

My name is Vitali

It was 6.00 p.m. on Monday evening and our translator, Mrs. Irina Vorobyova, was about to introduce me to the reserve's senior wolf research biologist, with whom I was to spend the week. Irina introduced me to the dark-haired gentleman, who smiled warmly and said "Hello........my name is Vitali". A warm handshake followed and Irina informed me that Vitali knew a little

English, but not much. However, we both had our English/Russian dictionaries in our hands, and I showed mine to Vitali, he in turn held his up to me, we laughed. I knew this was to be the start of a very special friendship.

Vitali Kochetkov is a quiet, intelligent man. When he speaks, he speaks with commitment, authority and understanding, especially about the subject of wolves. At that time, Vitali had been studying the wolves of Russia for over 17 years and he started to fascinate me as soon as I met him. Through Irina's initial translation, Vitali told me that he was delighted to meet with me, and he promised an exciting week ahead. It was going to be an important week as well, as a new wolf pack had recently denned just over six miles away from the village and it was this pack of nine wolves that presently interested Vitali.

Through our first hour's conversation, Vitali told me of a breath-taking predator prey scenario he had witnessed the year before, between four grey wolves, a cow elk and her calf. The four wolves had apparently cornered the cow, with her calf behind her, near a dense cluster of trees. The calf was injured and had difficulty walking, and the wolves had already sensed this during their initial stalking of their prey. Vitali then described, in great detail, the initial attack strategy of the four wolves and how the cow elk attempted to defend her calf, by lashing out at the wolves with her front and rear hooves. In a confrontation that lasted over three hours, one of the wolves was killed by a direct injury inflicted to the skull, and the cow elk also died alongside her calf. Only three wolves were left to the feast.

Later in the week, Vitali took me to the scene where this had occurred, and he showed me the left over carcass bones of the two elk and the wolf. I was amazed as Vitali pointed to several of the tree trunks around the area. Large hoof marks had been imprinted into the bark of the trees, where the cow elk had lashed out in all directions trying to protect her calf. This served to remind me, once again, just how dangerous the hunt can be for wolves.

Our initial introductory conversation ended as Vitali looked sharply at his watch and pointed to the figure nine. Irina told me, by this gesture, Vitali was indicating that, at nine o'clock, in just over one hour, he and his companion would pick me up and take me out for our first night's tracking of the nearby wolf pack.

"Be prepared..............for hard work" said Vitali.

Military manoeuvres

It was 9.00 p.m. on that beautiful Monday evening, and the old Land Rover came to a halt. I had been travelling with Vitali, Sasha, and two of their companions for nearly 30 minutes, heading north from the village into our 'study' wolf pack's territory. All four men were dressed in old, army style clothing which I found quite interesting at the time. I asked Vitali, whilst travelling, about their standard of dress. He explained to me that a great many of the park wardens were ex. Russian army personnel, who had decided to try to aid tourism around the village, by helping to track and to watch the wildlife of the outlying area. I found this aspect of conservation fascinating, and I applauded Vitali and his friends for their insight into both business and conservation.

As the Land Rover stopped, Sasha spoke to his friends in fluent Russian, then gestured that the three of us leave the vehicle. Sasha himself could be described as a gentle giant, and he too, like Vitali was a very sincere man with a great love and respect for wolves. We stood waving to our comrades in the vehicle, as they drove off, then Vitali started to thumb through his Russian/English dictionary. "Ian...." he said "we walk for.........one mile now.......to wolf habitat". I smiled and picked up my rucksack, "this was nothing like Minnesota", I thought, "but then again how could it have been?" I smiled at my friends, and then started the long walk. Over half an hour later, with much conversation about wolves having been thrown into the trek itself, we reached our destination. It was a large open field, with a beaver dam over on one side, clumps of logs on the other and across from us, half a mile away, a hidden wolf den site. "Now we eat" said Vitali.

That evening, as darkness crept in around us, we talked of many things, of wolves, our countries and peoples and, of our newfound friendship. As I sat eating my stale bread and chewing on goats milk cheese, Sasha kept smiling and talking to Vitali. Then Vitali translated to me, and vice-versa. It was certainly a three way conversation, with Vitali in the middle.

It was now 11.15 p.m., and as the blackness of the night engulfed us, the mood of my friends changed somewhat, and our voices had become whispers. "Ian....now we move.....wolf hunt soon". The adrenaline started to pump inside me as we gathered everything together and set off across the field. As I looked ahead, I noticed a long, straight line of grass that had been trampled down and, as Vitali noticed my interest, he started to thumb through his dictionary again. "Wolf puppy....." he whispered "hunt mice

and play together". I was elated, and both Vitali and Sasha knew it. As we walked on under the half moon, I could see the wolf pups in my own imagination as they gambolled through the grass. "This is going to be a splendid night" I thought.

Ten minutes later we had entered the main forested area, and I had to be especially careful where I trod. I strained my eyes in the dim light as various sounds of the forest assaulted my senses. At times, Vitali and Sasha would disappear from my side, only to turn up again a few minutes later, much to my shock and amusement. There was no doubt at all that these two men knew this part of the Taiga Forest extremely well.

Suddenly, Sasha stopped and put his left hand up to his ear. He turned to Vitali and I. Sasha had heard something which had obviously escaped me. Vitali crept up to me and whispered "listen.....". It was then that I heard the howling, the howling of wolves.

As we listened, the howling started to fade into the distance. Vitali, myself and Sasha were perched on an old, felled tree trunk where we had just prepared to start howling ourselves, in an effort to gain an approximate position of the pack, but the wolves had taken us by surprise. Sasha and Vitali looked at each other, then spoke in Russian. Vitali took me by the arm "we move with the wolves.......quickly" he said. We leapt off the tree trunk and I started to follow Sasha. By now the mosquitoes were terrible, and my vision had become hampered by the thick mosquito net over my head. As I scrambled over branches, twigs and undergrowth, I heard the wolves howling again, and this time the pups were yapping in unison with their older pack members. They stopped, and the forest fell silent. The three of us stopped too. The wolves were definitely on the move. We started again on the trail, with Sasha leading the way. It took well over an hour of concentrated effort to keep up with the wolf pack, which by now, had moved out to open grassland. Then, our efforts were rewarded, for there, right in front of us, was the pack's single telltale line of movement through the thick, damp grass. "Wolf move quickly" said Vitali, as Sasha started to imitate the call of a large owl that had just flown overhead. I watched in complete wonderment as the owl flew back to Sasha and then started to circle, approximately twenty feet above us. I was impressed, but then, it was back to business. "Wolves move pups.....move through village". Vitali was talking rapidly in hushed tones, as Sasha walked over to us. The wolves had indeed moved very swiftly and we had a lot of catching up to do. The pack itself had moved in the direction of one of the smaller villages, just as Vitali had predicted. It was obvious to the three of us, by this time, that the wolves were heading to a new rendezvous site, which turned out to be over two miles away on the opposite side of the village itself. We moved on.

Over an hour and I was exhausted. My feet were cut from the chaffing of the thigh length wellington boots, and I was hungry, but I was also incredibly happy. Across the horizon, as daylight just started to break, we heard the dogs from the nearby village barking wildly. The wolf pack had moved through the village, with no apparent concern regarding the humans or domestic dogs in the area. Apparently, as Vitali later told me, this is a common occurrence in the central forested areas of Russia. Wolves do move into villages, sometimes to take livestock such as chickens and goats, and sometimes just to scavenge scraps from the leftovers of village society. That particular night, they had taken nothing. All the wolves had wanted to do, was to move the pups to an area of newfound safety.

By 7 o'clock the next morning, we had finished our tracking, and Vitali had given an approximate location of the wolves' new living area. As we sipped our tea and started breakfast, Vitali, Sasha and I discussed the night's events. With Vitali and Sasha by my side, the three of us had tracked our wild canid friends to their new home. We had heard the wolves growling in the undergrowth and howling in the distance, and their paw prints had been all around us at one stage of the night's tracking. It had certainly been a night to remember.

At the end of the week, when I said my goodbyes to Vitali and Sasha, I felt a tinge of sadness. Our final handshakes were strong and warm, with a feeling of true friendship. "Sasha and I have enjoyed tracking wolves with you Ian" said Vitali. I returned the compliment and then boarded the bus which would take us back to the train station, for the long journey to Moscow.

That tinge of sadness grew heavier, as I waved to Vitali and Sasha from the back of the bus and, as the two of them became dots on the horizon, I settled into my seat and whispered "goodbye my friends......and thank you".

CHAPTER 5

Wolves and Ravens
Spirits of the Forest and Black Thunderbolts

The branches of the silver birch tree weighed heavy with the snow and with the small flock of large black birds. It was deep into winter in Southern British Columbia, Canada. The black birds screeched and cawed to each other, as movement down below caught their interest. The silver birch then let go of its avian friends. One of the birds flew zigzag through the snow blasted sky. The others enjoyed the show, and so too did the wolves........a large pack moving with speed through the thick snow, watching, waiting and following their winged colleagues. There was a scent. The largest wolf scanned the skies as the birds flew on, into the trees further ahead. Tails wagged and whines from the yearlings sent exciting messages to the rest of the pack. The wolves moved on.

The birds had done their job well. Quickly hopping down from branches to the snow covered floor of the forest, they jumped out of the way as the dog like carnivores moved past, panting heavily. Food was on the offering as the wolves moved around the old bull elk living its last few hours of life. The elk must have known the end was near. The wolves certainly knew it, and so too did the ravens.

The wolves were in no hurry as they settled down for a few hours wait. The ailing old elk looked around at the hunters he had kept at bay throughout his life. The snow kept falling........and the ravens kept calling.

These scenes are common in great wilderness areas of North America. Wolves and ravens have acted out their fragile alliance for thousands of years and each animal respects their fellow colleague, for where wolves still roam, the raven will always be there, following, listening, watching and calling.

The black thunderbolt of the sky

There are hundreds of thousands of birds on our planet. Some of these birds are noted for their plumage, some for their grace and beauty, some for their flying technique and some for their intelligence. Ornithologists deem the raven to be one of the most intelligent and distinctive of all avians. It is an imposing and mysterious bird, with its hoarse croaking call demanding attention and respect. Indeed, no other animal has such a great variety of vocal communication, except for humans. The raven is powerful and graceful and its flight patterns and techniques are quite remarkable as it swoops, rolls and dives through the sky.

Nicknamed 'the black thunderbolt' by many wildlife biologists, the common raven (Corvus corax) belongs to the avian family of the Corvidae. This class of birds includes rooks, crows and magpies, but it is the raven which stands supreme as the brains of the Corvidae world.

Named Corvus corax in 1758 by Carolus Linnaeus, the inventor of Latin classification of animal species, the raven inhabits all the main circumpolar regions of the world, including Europe, North America and parts of the Arctic Circle. Easily recognised by their large size, ravens generally weigh at least twice as much as the common crow, and their pointed wings, wedge-shaped tail and large bill give the birds an imposing look. Their wing span can average over four feet in length and they are highly intelligent and adaptable.

Wildlife research biologists have studied the interactions of wolves and ravens for many years, and have noted just how much both species have in common. Like wolves, ravens are monogamous, they are highly social and live in groups with a leader in charge, who will be the dominant bird. Ravens are omnivorous and will eat both animal and vegetable matter and they will hide surplus food, for when hard times lie ahead. Also, a most intriguing aspect of the bird's behaviour is how it will associate with humans where it is not persecuted.

A totem of respect

Just like the wolf, the common raven has attracted people's imaginations from the dawn of humankind's history gaining itself a vast array of interest in both folklore and in literature.

Hundreds of years ago, across the battlefields of Europe and Scandinavia, the raven's screeching call was an ominous sound to the soldiers of those dark days of humankind's history. The cry of the raven meant one thing.......death, for wherever warring armies went, the raven always followed. Just like the wolf, the raven has earned a fascinating place in the mythology of northern peoples. In the world of the Norsemen the raven was dedicated to Odin, god of battle and of the slain, and was the symbol on the Danish standard. Odin had two raven familiars, Hugin (Thought) and Munin

(Memory), who would perch on his shoulders and bring him news of the deeds of gods and of men. To the Scandinavians, wolves were the symbol of death and destruction and the darker side of life. At Odin's feet would sit two wolves, Geri and Freki, to whom he would feed scraps of meat, they symbolised Odin's savagery and cunning. All four animals always followed Odin into battle, and for a Norse warrior to see a wolf and a raven before going into battle, meant that he was assured of certain victory. Indeed, the name of Wolf-Hraben (Wolf-Raven), was the name of a brave and powerful warrior. To the people of the dark ages, the raven was the 'battle bird' and the wolf the 'battle dog'. Epic tales from those times such as Beowulf and the Battle of Brunanburh gave testament to the wolf and the raven, to their courage, strength and intelligence.

In Europe, hatred of wolves and ravens was prevalent, nurturing dark and irrational fears, and because of this, both animals suffered in consequence. Some of this hatred arose from the habit of wolves and ravens scavenging human corpses after war and plague.

Over in the New World of North America, however, it was a completely different story. Many of the native tribes of this beautiful continent held wolves and ravens in great respect. To the Native American, the raven was the 'creator', a folk hero and a jester, while the wolf was a spiritual brother to call upon during the hunt, to help and to protect the brave warriors. The Pacific coastal Native American tribes called the raven Yel because of its loud call. This, of course, has led to our own use of the word 'yell', meaning 'to shout'. The two main clans of the Northwest coast tribes also made special totems to their most respected and revered animal gods, namely the wolf and the raven.

The myths and legends surrounding the wolf and the raven across North America are far too numerous to recount in this chapter. It is, however, worth noting that the Eskimos of the great frozen northlands also held wolves and ravens in the greatest of respect, distinguishing them both as powerful gods and companions.

With the arrival of the white man into North America came the age-old hatred, persecution and suffering for both the wolf and the raven. In his book Hunting and Trading on the Great Plains : 1859 - 1875, James R. Mead comments on the deaths of thousands of ravens after they fed on carcasses laced with poison that had been left out for wolves. Because of the European settlers' gross ignorance and fear of the North American wilderness, both the wolf and the raven suffered great atrocities at the same time.

Nature's fragile alliance

Wolves need ravens and ravens need wolves. The two animals co-exist in what might be called a fragile alliance. In the world of the wolf and the raven, interaction is the name of the game and the final objective of this teamwork is to secure food.

Ravens are mainly scavengers and will eat mostly carrion. A flock of ravens can easily detect and catch sight of a dead animal such as a moose or a bear, but to enable the ravens to actually get into the carcass, they need the power and strength of the wolf. Likewise, a hungry wolf, or wolf pack, will eagerly follow a flock of ravens knowing what will be at the end of the trail.

Many wildlife research biologists have noted that, when following wolves, ravens will fly ahead of them, then let the wolves catch up, then fly on ahead again, as if following, but leading at the same time. Wildlife biologist Audrey J. Magoun often noted how a lone raven would follow her and her 'wolf-like' dog when they went for walks on the tundra. The raven would fly on, then wait for them to catch up, then would repeat the process again. Ms. Magoun noted in her studies that it was certainly advantageous for ravens to keep in close proximity to wolves, especially during the winter when daylight is so limited. Ravens will also fly excitedly over wolves on the hunt, anticipating the spoils to come, with their frenzied calls rising in volume at the scene of the kill itself.

When looking at the interaction between wolves and ravens, it seems that wolves are remarkably tolerant of ravens at the kill. However, there is always a small chance of wolves attacking ravens, as witnessed by biologists Durward L. Allen and Rolf Peterson on Isle Royale National Park in Lake Superior. This does, however, seem to be the exception rather than the rule, and especially if one particular raven appears to be too bold for its own good. One fact is certain though, ravens generally respect wolves at their meal times, although they may still try to harass the wolves from time to time. A common sight at most wolf kills in North America, is one or two cheeky ravens pecking the back legs of young wolves as if to say "it's our turn now".

Another interesting fact regarding the interplay between these two splendid animals, is the possibility that ravens have the ability to track wolves from the air. Again, researchers have noted that ravens can track wolves by following either their paw marks or tell tale 'line' of wolf pack movement through snow. This observation has led research biologists to imitate wolf howls to see, and make note, of how many ravens will appear in the trees around them at that time. One of many examples of this kind of interaction has been with Professor Fred Harrington, who has attracted ravens on many occasions whilst out conducting wolf howling surveys in parts of Canada.

One of the more unpleasant aspects of the raven's association with the wolf, is the fact that the birds are often seen consuming wolf scats. This happens particularly when food is scarce. Although this would be considered an extremely low form of nutrition, it still seems a viable energy source as far as ravens are concerned.

Play is also an important part of raven behaviour, just as it is with wolves, and one of the most beautiful and comical sights in nature is to see wolves and ravens playing 'raven tag'. The game starts with a lone raven teasing a wolf by calling to the wolf from just a few feet away. The wolf will then pounce at the raven, only to see the bird fly backwards and land a few feet away. Sometimes the raven will land on the wolf's back, only to fly off again as the wolf spins around, tail wagging, trying to catch the raven. Both animals seem to enjoy this game which is a delight to watch.

And so, the cycle between the wolf and the raven goes on to this day in North America's wild lands, and in many other parts of the Northern Hemisphere. The fragile alliance between Canis lupus and Corvus corax is a bonding of mutual tolerance, acceptance, trust and most importantly.........survival.

CHAPTER 6

The Red Wolf ~ Living On The Edge

A legend in the making

Just over thirty years ago, a copper red coloured wild dog of the southeastern United States, seemed doomed to total extinction.

The history of the red wolf and its fight for survival reads like a legend in the annals of wolf research. Yet, thankfully, today this beautiful animal is now living once again in its own natural habitat, after being declared biologically extinct back in 1980.

The saga of the red wolf recovery programme is both exciting and highly moving. It is also testament to human tolerance, care, understanding, and the founding of a major captive breeding programme. This is the story of the red wolf, and its legendary fight for survival.

Getting to know Canis rufus

It was Captain John Smith, who in the year 1624 described a "red-coloured wild canid, not much larger than our own fox", in his book A General History Of Virginia. This was the first time the red wolf had been described in literary history and to Captain Smith, at that time, this wild dog was most certainly a distinct species of wolf.

Many scientists and biologists now believe the red wolf hunted on the North American continent at least three quarter of a million years ago, and utilised a major section of the southeastern sector of the United States as habitat. Fossilised evidence now tells us that the red wolf's former range extended throughout the south-east itself, from the Atlantic and Gulf Coasts, north to the Ohio River Valley, through central Pennsylvania and west to certain locations in central Texas and south-eastern Missouri.

This evidence, in turn, has led to the red wolf being classed as a distinct separate species, and certainly a breed apart from the grey wolf (Canis lupus) and the coyote (Canis latrans). The red wolf's scientific Latin name is Canis rufus, and it is a highly recognisable wild dog. The average height for a red wolf is 26 inches from toes to shoulder, average length 4 to 5 feet from nose to tail tip and an average weight of between 45 to 80 pounds. Red wolves have a short to thick coat, dependent on the time of year. The colour of the coat can vary from a brown, buff colour to a dark, copper red with dark flecked guard hairs. Red wolves also have noticeably pointed, long ears, long legs and what can be described as a 'delicate' frame with a thin face and narrow snout. The general prey species for the red wolf are deer, racoons, rabbits, and various rodents. The red wolf's social behaviour is also noticeably similar to the grey wolf as they live in small packs in a claimed territory, have alpha leaders and their offspring disperse on a regular basis. However, red wolves do not hunt in large packs because of the size of their prey, they generally hunt in pairs or on an individual basis.

As with the grey wolf, the persecution of the red wolf began with the arrival of the European settlers and their trends for the domestication of livestock. With the arrival of these settlers came the same ideas of hostility and fear of wolves that had been prevalent all across Europe. Over a period of 200 years or so, the red wolf's habitat was systematically destroyed, and the red wolves themselves were shot, trapped and poisoned. Over this period of time, the future looked very bleak indeed for Canis rufus.

In 1962, the American scientific community determined that the last remaining red wolves were now on the verge of extinction and in 1967 the red wolf was declared an endangered species by the Fish and Wildlife Service. This declaration was based upon the Endangered Species Preservation Act of 1966, but these were just words on paper, in the wild the persecution still continued. It was in 1973, however, when U.S. Congress passed 'The Endangered Species Act' that the red wolf was immediately listed and biologists began realising the terrible truth of what was happening to the last remaining red wolves. By this time it had become a well-known fact that the last red wolves were actually hybridising with coyotes. Considering that, by 1970, the entire population of Canis rufus was believed to be less than 100 animals, confined to a small area of coastal Texas and Louisiana, the biologists then realised that something had to be done, and quickly, as the last red wolves were then living on the edge................of extinction.

A race against time

Between 1974 and 1979 an all-out effort was made to trap and collect the last remaining wild red wolves. It was a race against time because of the coyote hybridisation factor. Trappers tried their best during the capture period to keep a 'no coyote' buffer zone in the red wolf area, but it was very difficult. Gut-wrenching decisions had to be made by the U.S. Fish and Wildlife officials as the remaining captured red wolves were evaluated for purity, before being placed into the newly formed captive breeding centre at Point Defiance Zoo in Tacoma, Washington State. At the time of the capture process, it was found that animals seeming like pure red wolves on the outside, had coyote blood on the inside. The breeding programme could only be utilised from pure red wolf genes, and eventually through captive breeding experiments, only 14 captured animals were deemed pure enough to form the nucleus of the red wolf recovery programme.

At the start of the programme, Point Defiance Zoo was well-known for its interest in preserving endangered species and had already been established as the first centre for the hopeful captive breeding of red wolves since 1969. The first major success for the zoo came in 1977 when the first captive red wolf pups were born. By the end of 1978 the captive population stood at 11 males and 7 females, comprising of more newly born pups and newly captured pure red wolves. An 'experimental' release of red wolves also took place in 1976 and 1978 at Bulls Island off South Carolina, where the red wolves were initially released, tracked and then re-captured. Now, at last, the future for Canis rufus was starting to look a little bit brighter. In 1979, the last remaining pure red wolves were captured and taken to Point Defiance Zoo where only one was determined to be a pure red wolf. Canis rufus, as a species, was now in total captivity.

In 1980, the red wolf was officially classed as biologically extinct in its own natural habitat.

A decade of success

Due to the success of the live tracking and pre-release on Bulls Island, in October 1980, another breeding facility was established, this time at the Audobon Zoological Park in New Orleans. At that time, the captive population of red wolves was increasing even further with 21 males and 16 females in holding pens. Between 1981 and 1984 the breeding programme grew again, with the participation of Texas Zoo and Alexandria Zoo in Louisiana. It was, however, in 1984 that a real breakthrough happened which came from a donation of a large piece of land totalling 120,000 acres, by the Prudential Insurance Company. This piece of land was known as the Alligator River National Wildlife Refuge (ARNWR) and it was to play a key role in future red wolf recovery. The ARNWR was large enough to support several small wolf

packs and contained enough small mammals for the wolves to prey on. It was also sparsely populated by humans, and had no coyote populations in the area. At last! A major release site had been found, thanks to human understanding and generosity.

Plans were then immediately put into action and wildlife research biologists worked for three years on an in-depth plan for the first major release of red wolves into the wild. Thanks to the great success of grey wolf recovery in Minnesota at the same time, a great deal of knowledge had already been acquired by the biologists, especially regarding the use of radio collars and telemetry tracking. The few residents in the ARNWR area were also notified of a forthcoming release, and were assured that no wolves would stray from the refuge, thanks to the use of the ingenious capture collars.

Nevertheless, the biologists still had reservations about a forthcoming release as many questions remained unanswered. How would red wolves survive for long periods of time? Had their hunting instinct survived, and would pack mentality be the same in the wild after years of captivity? To the research biologists, there was only one way to find the answers.........

On September the 14th 1987, the first pair of captive-bred red wolves were fully released in the ARNWR. Meanwhile, at the same time, another pair of red wolves had been shipped to Bulls Island to fully establish an island propagation project, to help red wolves acclimatise to a full release into a major land mass area. At the end of 1987, the captive population then stood at 34 males and 46 females. The red wolf recovery programme was, at last, gaining momentum.

On April the 28th 1988, the first red wolf pups were born in the wild in the ARNWR. History had been made. By the end of 1988 the captive population stood at 36 males and 47 females. In 1989 red wolves had also been released onto Horn Island, off the coast of Mississippi, to help initiate a second island propagation project. By the end of 1989, the total population of captive wolves stood at 45 males and 60 females. The red wolves had responded well, both in captivity and in the wild. Canis rufus had returned from extinction thanks to a decade of success.

Into the nineties............Survival of the fittest

Between 1987 and 1992, a total of 36 captive bred red wolves had been released in the ARNWR. The mortality factor for these wolves was quite high, with 21 of the 36 being killed either by drowning in the swampy wilderness, or by territorial conflicts, but this was to be expected. Another 7 wolves had also been recaptured because of their tendency to stray out of the refuge itself, proving the worthiness of radio collaring. Meanwhile, in 1990, several more captive-bred red wolves had been released on St. Vincent Island off the coast of Florida. This was to become the third propagation project, and proved highly successful. In 1991, red wolves were released into Tennessee's Great Smoky Mountains National Park (GSMNP). This huge, beautiful area, consisting of 520,000 acres was to prove less successful in red wolf recovery. By the end of 1992, the red wolf population in the ARNWR was starting to increase, with the second generation of red wolf pups being born, and more captive wolves being shipped to various zoos across the United States, as part of an expansion programme of captive breeding.

By the end of 1993, another great wilderness area had been expanded into the programme, this was the 113,000 acre Pocosin Lakes National Wildlife Refuge, and in that same year, the first wild red wolf pups were born in the GSMNP. Throughout 1994 to 1997, the red wolf population, both wild and captive increased and decreased with a natural flow, as was to be expected with Mother Nature's rules of survival of the fittest.

Now, as we reach the new millennium, the red wolf population seems to be doing well. The actual target for recovery, as set by the U.S. Fish & Wildlife Service is 550 red wolves in total, consisting of at least 3 wild populations totalling 220, and 330 wolves in captivity, throughout 30 or more facilities. At the end of August 1997, the total population was 240 to 317 wolves, both captive and wild, so the programme still has a long way to go. However, success is on the horizon, although in October 1998 the U.S. Fish & Wildlife Service announced the cancellation of the GSMNP recovery programme. Sadly, this project had failed due to the high mortality rates of red wolf puppies in this area. Over a period of time, research biologists had discovered various problems were occurring, hampering the growth of the pups. These problems included disease, predation, malnutrition and parasitic infection. In October 1998, it was then agreed that the last remaining red wolves in the GSMNP were to be recaptured and later released into another recovery area sometime in the future.

One fact is certain, however, through human understanding, care and tolerance, one of this planets most endangered wild dogs now has a chance of survival, and the howl of the red wolf can still be heard in certain wild places, in the south-eastern United States.

CHAPTER 7

Man's Best Friend.......The Wolf

The Superior National Forest........Northern Minnesota.......The small herd of white-tailed deer were nervous. Something was alive in the forest, and moving towards them. The small ungulate's ears were twitching rapidly, detecting sounds we humans could never hear. Then....a snap of a twig....a short, deep growl.....and the wolf pack attacked.

Fleeing in all directions, the deer moved rapidly, but one of the older members of the herd, was not so nimble. Old age had caught up with it, and the wolves knew. The large wild dogs moved with determination towards their prey. This time, success would come the wolf pack's way. Within minutes the deer had fallen. With the death of one animal, came the survival of others.

Meanwhile, several thousand miles away, in a field in the county of Cumbria, in England, a lone farmer leans on a gatepost....smiling. Then he gives a whistle and a shout, for his three border collies have amazed him once again. The sheep are gathered in their pen, and the farmer's three loyal friends sit on the grass.....panting. They had all worked well together for many months, and the farmer always greatly appreciated the co-operation and unique strategy and intelligence that his dogs showed. But, what the farmer loved most of all about his dogs, was that, at times, they seemed so wild....and so free. Utilising all their senses, in so many ways whilst rounding up the sheep, the border collies certainly behaved and acted, at times, like wolves. Of that fact, the farmer was certain.

A wolf in dog's clothing

If you live with a dog, then you are actually living with a domesticated wolf, and that's a fact! Through thousands of years of domestication, all our dogs have become mutations of the world's largest dog, namely Canis lupus.....the grey wolf.

People from all walks of life have long been fascinated with the relationship between our dogs and wolves. Indeed, over in Los Angeles in the United States, a major scientific study has recently taken place under the guidance of research scientist Bob Wayne, utilising DNA samples of many types of wild dog and comparing these samples with the DNA of our own domestic dogs.

The only wild dog DNA sequences to be confirmed, within all the domestic dogs in these tests, was that of the grey wolf. So, it is now official, whether you have a great dane, a poodle, a husky, a corgi, or any other breed of domestic dog, you are actually living with a wolf......in dog's clothing.

There are two major theories as to when the domestication of our dogs, from wolves, actually took place. Some scientists state that domestication probably happened between 10,000 to 20,000 years ago, as fossilised evidence of early domestic dogs has been traced to around this time. However, scientists in Los Angeles are now stating that the domestication of wolves possibly began as early as 135,000 years ago, as fossilised wolf bones have been found in or around early human settlements of that time. This greatly suggests that our first 'best friends' were actually........wolves.

The actual 'genetic' theories regarding these studies are far too complicated to cover in depth here. However, these theories can be explored further by referring to the references listed at the end of this book.

One fact is absolutely certain though, domestication of the grey wolf did take place at sometime in early human history. Thousands of years ago wolves did sit around our campfires, possibly expecting the last morsels and bones from the human feast and, as humankind accepted these wild dogs into their settlements, they became tame.

Your dog's behaviour......Is wolf behaviour

Have you ever wondered why your dog licks your face in greeting, or rolls on its back to have its stomach stroked, or why it will stop suddenly just to sniff an area, then urinate on that area?

These three manneristic traits along with many other behavioural characteristics, show all the same types of behaviour, as those shown by grey wolves. The behaviour of our domestic dogs is yet another piece of evidence which proves their relationship with Canis lupus.

I have listed below a set of behavioural patterns, as shown by both wolves and dogs, and an explanation for each. Some of these are already mentioned in chapter one, but it is important that we cover these aspects again for this chapter as follows:

All wolves are highly social, intelligent animals and live in packs. The wolf pack itself is an extended family unit consisting of the dominant male and female, and various family members of varying ages. Your dog views you as a pack member, and is happy to be a member of your 'pack', just like a wolf.

The dominant male and female in a wolf pack are called the 'alphas'. Generally, they are the only breeding pair within the pack itself. All other pack members will show submission and respect to the alphas at all times. Your dog will also show you respect and submission by ranking you as the alpha leader of your 'supposed' pack.

Wolves show two basic types of submissive behaviour to the alphas. These are known as 'active' and 'passive' submission. Active submission is much in evidence during the greeting stage of behaviour, or before, or after, a hunt. This will involve much muzzle licking, tail wagging and general excitement towards the alpha animals. Passive submission comes from a dominant form of behaviour by the alphas towards a 'given' wolf in the pack. Most probably, this will be the lowest ranking pack member, called the 'omega' wolf. This chosen wolf will then probably lie on its back, with legs and underbelly 'open' to the alphas. This shows that no threat whatsoever is posed to the pack leaders. Your dog will show exactly the same types of active and passive submission to you as the human pack leader, hence examples of 'face licking' and 'tummy rubbing'.

As far as I am concerned, however, based on my own years of researching wolf/dog behaviour, I believe that a third type of submissive behavioural pattern is evident. I call this 'aggressive submission'.

This type of submissive behaviour in wolves is seen at the kill. On rare occasions, a lower ranking wolf may approach the carcass, before, or with, the alphas. This wolf, at this time, will show signs of fear such as curling its tail in between its hind legs and arching its back downwards as it 'crawls' towards the food source. However, at the same time, this wolf will also show aggressive signs such as baring its teeth, snarling and growling. In general, this behaviour will only last a few seconds before the alphas will 'punish' the wolf for lack of respect and for stepping out of the heirachial line. This, to me, is aggressive submission.

Your dog will sometimes show exactly the same behaviour over its food. How many times has your dog snarled, in a submissive way, when you have approached its meal? At this time, your dog knows that you are the alpha animal, but it still feels the urge to show aggressive submission over securing its own food source.

To both wolves and dogs, these traits of submission are very important aspects of behaviour, whether they be active, passive or aggressive.

The licking of a muzzle is also a sign to regurgitate half-digested meat from within an adult wolf's stomach, especially to feed the pups. Wolves never really grow out of this general greeting behaviour, adopted from puppyhood. Your dog licks your face in exactly the same way, feeling a compulsion to beg for food during the greeting process. It is also a submissive gesture towards a dominant animal.

Communication is a vital aspect in both wolf and dog society. Wolves will bark, whine, winge and grunt at times, and body language also plays an essential role in communication. The howling of wolves is also vital for social interaction, defensive strategy and motivation. How many times have you heard your dog howling to certain sounds such as music, a siren or an alarm? Your dog, on occasions, still feels the 'call of the wild' and will respond in its own lupine way.

In the wild, wolves claim territory by scent marking, using glands in both their anal area and their paws. The main type of scent marking is called 'raised leg urination', your dog does exactly the same to lay out its own territorial boundaries, and to remind other dogs in the area just who claims this set piece of territory.

Before the hunt, wolves will do their best to cover their own scent by, what wildlife biologists call, 'olfactory sensing'. This is more commonly known as 'scent rolling'. This trait of behaviour allows wolves to get close to their prey, without their own scent being detected. In general, carcasses of other animals or pungent sources on the ground will be used. The wolf, or wolves will then begin their role by probably bending their forelegs then rubbing the side of the neck and temple area on the object, several times, on alternating sides. This may then be followed by the animal turning completely over onto its back and wriggling on the object making sure its main back area is covered with the new scent. Many breeds of domestic dog will, at times, feel a compulsion to do exactly the same, much to the consternation of their owners.

Success on the hunt brings survival for all wolves. Indeed, the wolf itself is considered a top predator in the Northern Hemisphere. All our domestic dogs still, in many ways, have the desire to hunt. A friend of mine has a German shepherd dog which thoroughly enjoys hunting, killing and then eating, rabbits. If you ceased feeding your dog and left it to look after itself, it would survive by scavenging......and hunting. As far as your dog's survival was then concerned, the carnivory aspects of the wolf and strategies for the hunt, would soon become established or the dog would perish.

There is no finer example of hunting strategy and tactics in the domesticated dog's world, than to watch border collies rounding up sheep. Although the farmer gives the commands, these beautiful, intelligent dogs show a splendid degree of pack cohesion and co-operation, utilising several stages of the hunt itself, namely....locating of the prey, stalking of the prey, the confrontation and the chase. Just like wolves hunting caribou or bison, the collies, and other dogs, will utilise their intelligence on a domesticated basis, for the hunt.

Dogs will feel a compulsion, at times, to bury their favourite bones, or possibly, a piece of meat. This comes from a trait of behaviour in the wolf called 'caching' of meat. Wolves will bury surplus meat from the kill and store it for harder times ahead, especially in areas of 'low prey density'. Cached meat brings survival in the wild. Your dog is behaving just like a wolf when it behaves in this way.

Just like wolves, dogs will also defend territory when they feel the need. At times wolf packs will defend their territory to the death against other wolf packs and carnivores, especially in defence of their pups. Postmen and milkmen have often become 'victims' of 'territoriality' in the domestic dog's world. Your actual home or living area is defined as territory by your dog, and being a faithful pack member it will show aggression to any unwanted intruder.

Play, of course, is vital for any developing puppy, both domestic or wild. Your puppy, or puppies, will show all the fun, mischief and excitement as that shown by grey wolf pups when they start to leave the den. Over the coming months wolf pups will become extremely inquisitive regarding all aspects of the outside world. They will play fight and explore, and will sleep often. Playing is an essential role in the wild, not just for the fun aspect, but to hone the skills that will be required by the pups once they are older. These skills will bring about survival in a tough environment, and will also mould new alpha leaders. Your dog puppies feel the same compulsion to explore, chew, fight and hunt, just as wolf pups do in the wild.

Play is also an important aspect of socialisation for yearling and adult wolves. The classic 'play bow' position, eyes wide, front legs and paws stretched out on the ground, ears back, backside and legs raised up and tail wagging furiously are all signs of "I'm having a great time, come and play with me". Your adult dog will still enjoy playing with you for many years, as part of the social cohesion of 'pack' mentality.

Why do domestic dogs make a fuss over their sleeping area? Why does your dog move around in a tight circle or 'rough up' a blanket before going to sleep? This is because all wolves make themselves as comfortable as possible in the wild, before sleep. A bed of compacted snow or gathered leaves makes a perfect resting place for a tired grey wolf.....or a domestic dog.

Finally, when we look at the behavioural aspects of wolves and dogs, we must also cover the utilisation of predatory senses. Your dog's senses are vastly superior to ours in many ways, this is because they have evolved from predators. Be assured that your dog has not lost any of its keen predatory senses. This is why domestic dogs are so successful when utilised in search and rescue, customs and excise, police duties etc.

All of these behavioural aspects of wolves and dogs, to me, prove one thing. Our dogs are most certainly domesticated wolves. They still have a feeling of wildness, and of wilderness deep down inside their genetic makeup. So, the next time you make a big fuss of your best friend, look deep into his, or her eyes, and I guarantee, you will see man's best friend.......the wolf.

CONCLUSION

Wolves, humans........and the future.

As we now come to the end of our journey into the world of the wolf, let us ask ourselves a very basic question.

"Will wolves ever become extinct?"

My answer is............no.

Thankfully, there are still thousands of wolves inhabiting the Northern Hemisphere. Canada alone has, quite possibly, the world's largest wolf population with an average total of approximately 50,000 plus. Wolves are now also starting to return to the lower 48 states, both through natural colonisation and man-made reintroduction. The states of Montana, Wyoming, Idaho, Wisconsin, Michigan and Arizona have all seen the return of wolves, and Minnesota's wolf population is still increasing rapidly.

Europe is also seeing a steady increase in wolf populations, with Italy being especially successful in wolf conservation and population management. Spain and Poland's wolf populations are still stable, and Germany has given protection to wolf packs in various areas, such as Brandenburg. France too, has seen an influx of migrating wolves, with Scandinavia witnessing a small increase in its population also.

Wolves still inhabit areas practically devoid of human settlement and interference. The colonies that were once part of the old Soviet Union still have thousands of wolves between them and, up in the far north, such as Alaska and Ellesmere Island, wolf populations are stable.

Obviously though, there will still be wolf/human conflict and, at times, humans will wage war on the wolf in given circumstances. Generally, however, people are now becoming much more interested in wolves and in the whole of nature itself. From this interest comes understanding and, most importantly......tolerance.

With this in mind, we must never forget that education is of paramount importance to help us understand the grey wolf and its role in the wild. Now that wolves are returning to their former homelands, education will be the key to helping ensure their survival for many years to come.

In conclusion, the fact remains that most people will never get the chance to see a wild grey wolf in its own natural habitat, but isn't it wonderful to know that they are still out there.

Bibliography - Suggested Reading

Adams L.G./Dale B.W./Mech L.D. 1995
Wolf Predation on Caribou Calves in Denali National Park, Alaska In: Ecology and Conservation of Wolves in a Changing World Ed: Carbyn/Fritts/Seip
Canadian Circumpolar Institute

Allen D.L. 1993 Edition
Wolves of Minong : Isle Royale's Wild Community
University of Michigan Press 19931

Berg B. 1994
Wolves In Minnesota
DNR Forest Wildlife and Research Group

Cohn J.P. May 1987
Red Wolf in the Wilderness
Bioscience Magazine

Carbyn L.N.
Grey Wolf And Red Wolf In: Wild Furbearer Management and Conservation in North America
Ontario Trappers Association

Carbyn L.N., Fritts S.H. & Seip D.R. 1995
Ecology and Conservation of Wolves in a Changing World
Canadian Circumpolar Institute - Occasional publication no. 35

Carbyn L.N./Oosenburg S.M./Anions D.W. 1993
Wolves, Bison....and the Dynamics Related to the Peace-Athabasca Delta in Wood Buffalo National Park
Canadian Circumpolar Institute Research Series No. 4

Carbyn L.N./Trottier T. Dec. 1988
Descriptions of Wolf Attacks on Bison Calves in Wood Buffalo National Park
Journal of the Arctic Institute of North America - Volume 41, no. 4

Endangered Red Wolves 1997
U.S.F.W.S. Special Edition

Fanshawe J. April 1988
Fidos Forefathers
BBC Wildlife Magazine

Fiennes R. 1976
The Order Of Wolves
Hamish Hamilton

Fox M.W. 1987
Behaviour of Wolves, Dogs and Related Canids
Jonathan Cape Press

Fogle B. Dr. 1992
Know Your Dog
Dorling Kindersley Ltd.

Fritts S.H.   Spring 1993
Livestock Depredation.......The Downside of Wolf Recovery
International Wolf Magazine

Fritts S.H.   1983
Record dispersal by a wolf from Minnesota
Journal Of Mammology No. 64

Fritts S.H./Paul W.J./Mech L.D./Scott D.P.   1992
Trends and Management of Wolf/Livestock Conflicts in Minnesota
U.S.F.W.S. Resource Publication No. 181

Fritts S.H./Bangs E.E./Fontaine J.A./Johnson M.R./Phillips M.K./Koch E.D./Gunson J.R.   1997
Planning and implementing a reintroduction of wolves to Yellowstone National Park and central Idaho
Restoration Ecology No. 5 (1):7-27

Greeley M.   1996
Wolf
Metro Books

Grooms S.   1993
The Enigmatic Red Wolf From: The Return of the Wolf
Northword Press

Harrington F.H./Mech L.D.   1983
Wolf pack spacing: Howling as a territory independent spacing mechanism in a territorial population.
Behavioural Ecology and Sociobiology Magazine

Hawkes N.   Friday, June 13th 1997
Genetics Exposes the Wild Past of Man's Best Friend
The Times Newspaper

Hayes B.   Summer 1993
Hunters of Moose
International Wolf Magazine

Hayes R.D./Baer A.   1992
Brown bear (Ursus arctos), preying upon grey wolf (Canis lupus) pups at a wolf den
Canadian Field Naturalist Vol. 106

Hayes R.D./Mossop D.H.   1987
Interactions of wolves (Canis lupus) and brown bears (Ursus arctos) at a wolf den in the Northern Yukon
Canadian Field Naturalist Vol. 101

Heinrich B.   1991
Ravens in Winter
Vintage Books

Heinrich B.   Sept/Oct 1994
Riddle of the Ravens
Wildlife Conservation Magazine

Horan J.   May/June 1986
The Red Wolf is Coming Home
Defenders Magazine

Hutchinson J.   Winter 1994
Strife Among Wolves: When Aliens Meet
International Wolf Magazine

International Wolf Magazine
Spring 1999
Red Wolf restoration halted in Great Smoky Mountains National Park

Judd C.D.   March/April 1986
Against the Odds
Animals Magazine (USA)

Kidder C.   Sept/Oct 1992
Return of the Red Wolf
Nature Conservancy Magazine

Long K.   1996
Wolves:
A Wildlife Handbook
Johnson Nature Series

Lopez B.H.   1978
Of Wolves and Men
Macmillan Publishing

Lawrence R. D.   1993
Trail of the Wolf
Key Porter Books

Massey-Stewart J.   1992
The Nature Of Russia
Boxtree Limited

McCrackan-Peck R.   1990
Land of the Eagle
Guild Publishing

McIntyre R.   1993
A Society of Wolves: National Parks and the Battle Over the Wolf
Voyageur Press

Mech L.D.   1991
The Way of the Wolf
Swan Hill Press

Mech L.D.   1981
The Wolf, the Ecology and Behaviour of an Endangered Species
University of Minnesota Press

Mech L.D.   Sept/Oct 1988
The Great Musk Ox Hunt
Defenders Magazine

Mech L.D.   Nov/Dec 1989
Stubborn Hunter in a Harsh Land
Defenders Magazine

Mech L.D.   1966
The Wolves of Isle Royale
US National Parks Fauna Series No. 7

Mech L.D.   1988
Longevity in Wild Wolves
Journal of Mammology No. 69

Mech L.D./Meier T.J./Burch J.W./ Adams L.G.   1995
Patterns of Prey Selection by Wolves in Denali National Park, Alaska In: Ecology and Conservation of Wolves in a Changing World   Ed: Carbyn/Fritts/Seip
Canadian Circumpolar Institute

Mech L.D./Adams L.G./Meier T.J./ Burch J.W./ Dale B.W.   1998
The Wolves of Denali
University of Minnesota Press

Morell V.   June 1997
The Origin of Dogs : Running With the Wolves
Science Magazine (Vol. 276)

Multiple and Ancient Origins of the Domestic Dog   June 1997
Various Authors
Science Magazine (Vol. 276)

Nelson M.E./Mech L.D.   1984
Observation of a swimming wolf killing a swimming deer
Journal of Mammology no. 65

Nowak   R. M.   1995
Another look at wolf taxonomy In: Ecology and Conservation of Wolves in a Changing World Ed: Carbyn/Fritts/Seip
Canadian Circumpolar Institute

Ovsyanikov N./Bibikov D.I./Bologov V.V.   Spring 1998
Battling With Wolves: Russia's Decades Old Struggle To Manage Its Fluctuating Wolf Population
International Wolf Magazine

Peterson R.O.   1977
Wolf Ecology and Prey Relationships on Isle Royale
National Park Scientific Monograph

Wolf Conservation And Education In The British Isles

WOLF HELP

We are a specialist, non-membership, educational/conservation team. We are supported by and have photographic slides from many of the world's foremost wolf biologists. Our lectures include up-to-the-minute information on wolf recovery and re-introduction, distribution, conservation, wolf society, behaviour and habitats.

The Wolf Help team, Ian Redman and Carole and Niel Stevenson, are happy to travel across Great Britain to give their presentations to interested groups of people, including zoos, colleges, natural history groups, schools and various professional animal/wildlife organisations.

Wolf Help can be contacted through the printers
or directly by e-mail:
wolf.help@tinyonline.co.uk

Ian Redman has researched, written and presents all of the following Wolf Help lectures:

Secret World ~ Discovering the truth about the wolf
A detailed look into the world of wolves and wolf society in general, including sub-species, mating, raising of pups, alphas to omegas, pack mentality and behaviour, hunting technique, territories, communication, wolf populations around the world, wolves and man, legends and wolf conservation. Including a detailed insight into wolf society in North America, man-made reintroduction, natural recolonisation, radio collaring and telemetry and wolf recovery in Europe.
Duration - 1 hour 30 minutes

Wolves ~ Spirits of the forest
General information about wolves and wolf society as in Secret World, but more emphasis on Native American Indian legend, humankind's effect on the wolf and modern day wolf recovery activities in certain parts of the world.

Also featuring the work of wolf biologists Dr. L. David Mech and Dr. Ludwig Carbyn in their respective areas of wolf research.
Duration - 1 hour

The incredible world of wolf research
Join eight of the world's finest wolf biologists in different locations throughout North America's wolf country. An exciting and thought provoking lecture taking the audience into the 'front line' of wolf research and conservation. A highly exciting visual presentation, including:-
- Diane Boyd - The wolves of Glacier National Park.
- Steven Fritts, L. David Mech, Jeff Haas - Operation Wolfstock, the reintroduction of grey wolves to Yellowstone National Park and Central Idaho (1995 - 1996).
- Will Waddell, Jennifer Gilbreath - The Red Wolf Recovery Programme.
- Adrian Wydeven - Wisconsin's Wolf Recovery Programme.
- Jim Hammill - The return of grey wolves to Michigan's Upper Peninsula.

(Also including two rare pieces of wolf research video footage).
Duration - 1 hour 15 minutes

Savage Freedom. . . . . .Wolves on the hunt
Enter the world of the predator and the prey in this hard-hitting, exciting and highly factual talk. Learn how grey wolves use their intelligence to hunt various prey species across North America. Discover how musk ox, caribou, elk, moose, dall sheep, bison and white-tailed deer defend themselves against wolf pack attack. Discover the role of the wolf in its regulation of prey animals in the wild, and of its use of prey search imagery. Including superb visual images of wild wolves and their prey species in Minnesota, Alaska, the High Arctic, Wood Buffalo National Park and Isle Royale National Park.
Enter the realm of grey wolves on the hunt................and discover a world of savage freedom.
Duration 1 hour

Man's best friend ~ The wolf
A fascinating view into the world of our own domestic dogs through the eyes of a wolf. Also, an in depth view of dog behaviour compared to wolf behaviour and how human beings are actually living with the domesticated version of the wolf.
Duration - 1 hour

Wolfways
An interesting and thought provoking view of the world of wolves and humans from the beginnings of mythology to modern technological wolf research

methods. Covering various aspects of the grey wolf from both European and North American viewpoints. From werewolves to radio collars, this presentation has something for anyone interested in wolves.
Duration - 1 hour

Back to the wild......the Mexican wolf and the Red wolf recovery programmes

Enter the world of North America's most endangered wild dogs. Utilising fully updated information and superb photographic slides, this presentation is fully endorsed by the U.S. Fish and Wildlife Service's leading biologists involved with these recovery programmes. Exciting and highly moving.
Duration - 1 hour

Closing the door on Little Red Riding Hood

Our ideas and perceptions about the wolf and how people are discovering the truth about wolves and their place in the natural world.
Duration - 50 minutes

Making friends with wolves (*for schools*)

One for the kids, especially 5 - 10 year olds. Plenty of chances for children to interact and lots of wolf pup slides and photographs of people enjoying being in the company of wolves. Lots of fun and a talk that children will always remember.
Duration - 50 minutes

The wolves of Denali

Utilising 100 photographic slides, most of which have been kindly donated by Dr. L. David Mech and Mr. Rick McIntyre, this presentation gives an in-depth view into the world of predator/prey dynamics and a history of wolf research around the Mount McKinley area in Denali National Park.
Duration - 60 minutes

International Wolf Conservation And Education

For anyone wishing to further their interest in both grey and red wolves, for a worldwide view of wolf conservation, I highly recommend contacting the following organisations:

The International Wolf Center
1396 Highway 169
Ely
Minnesota 55731
U.S.A.

Website address: http://www.wolf.org

The IWC has various membership rates, for both single members and families, which include an excellent quarterly magazine entitled 'International Wolf'.

Wolf Haven International
3111 Offut Lake Road
Tenino
Washington 98589
U.S.A.

Website address: http://www.wildwolf.org/

Wolf Haven also has a selection of membership rates and publishes a newsletter entitled 'Wolftracks'.

Mission: Wolf
P.O. Box 211
Silver Cliff
Colorado 81249
U.S.A.

Website address: http://www.indra.com/fallline/mw/

Mission: Wolf operates various 'Wolf Caretaker' rates and publishes a bi-yearly newsletter entitled 'Wolf Vision'.

ABOUT THE AUTHOR

Ian Redman is one of the co-founders of WOLF HELP, a British team who are, very successfully, educating the public about the nature of wolves and wolf society. Ian is the speaker for the group, he writes and researches all of his WOLF HELP lectures and maintains regular contact with North America's foremost wolf biologists who have constantly supported Ian's efforts on behalf of wolf conservation.

Although Ian has a full-time career in the food industry, he devotes all of his free time to his wolf studies. He is absolutely committed to wolf education and conservation and has an impressive library of wolf literature and biological papers at his studio style home in Northwich, Cheshire.

Ian has had several wolf articles published both in Britain and abroad and attributes the beginnings of his fascination with the wolf to a documentary he watched on TV in 1989 entitled The Ice Pack, about the study of Arctic wolves on Ellesmere Island, featuring Dr. L. David Mech and Jim Brandenburg.

Since 1989, Ian has tracked wild wolves in the Taiga Forest in Russia and, on two occasions, in Minnesota. He has taken an active part in the Mexican Wolf Recovery Project when he was invited to Wolf Haven International in Washington State where he assisted with the capture and veterinary care of eleven captive-bred Mexican wolves, several of which were later released into the wild in Arizona.

This is Ian's first book.